The Business Manager's Guide to Software Projects

A Framework for Decision-Making, Team Collaboration, and Effectiveness

Jonathan Peter Crosby

Apress®

The Business Manager's Guide to Software Projects: A Framework for Decision-Making, Team Collaboration, and Effectiveness

Jonathan Peter Crosby
Baden, Switzerland

ISBN-13 (pbk): 978-1-4842-9230-3
https://doi.org/10.1007/978-1-4842-9231-0

ISBN-13 (electronic): 978-1-4842-9231-0

Managing Director, Apress Media LLC: Welmoed Spahr
Acquisitions Editor: Joan Murray
Development Editor: Laura Berendson
Editorial Assistant: Gryffin Winkler

Cover image designed by Isaac Soler at eStudioCalamar.

Distributed to the book trade worldwide by Springer Science+Business Media New York, 233 Spring Street, 6th Floor, New York, NY 10013. Phone 1-800-SPRINGER, fax (201) 348-4505, e-mail orders-ny@springer-sbm.com, or visit www.springeronline.com. Apress Media, LLC is a California LLC and the sole member (owner) is Springer Science + Business Media Finance Inc (SSBM Finance Inc). SSBM Finance Inc is a **Delaware** corporation.

For information on translations, please e-mail booktranslations@springernature.com; for reprint, paperback, or audio rights, please e-mail bookpermissions@springernature.com.

Apress titles may be purchased in bulk for academic, corporate, or promotional use. eBook versions and licenses are also available for most titles. For more information, reference our Print and eBook Bulk Sales web page at http://www.apress.com/bulk-sales.

Any source code or other supplementary material referenced by the author in this book is available to readers on GitHub.

Printed on acid-free paper

Table of Contents

About the Author

Jonathan Peter Crosby is a software developer, performance engineer, and consultant who has worked in the field for over 20 years.

Having gained his professional experience at a range of companies from start-ups to blue chips, he also founded and cofounded three small tech companies—the first one at the age of 33. Jonathan works at the crossroads of technology and business. The best project outcome, he finds, is achieved through sharing the business and technical knowledge—a reason that he likes to involve all team members in important project decisions. Jonathan believes that effective communication is the cornerstone of every successful software project.

About the Technical Reviewer

 **Juval Löwy** is the founder of IDesign and a master software architect. Over the past 20 years, Juval has led the industry with some of his ideas such as microservices serving as the foundation of software design and development. In his master classes, Juval has mentored thousands of architects across the globe, sharing his insights, techniques, and breakthroughs, and has helped hundreds of companies meet their commitments. Juval participated in the Microsoft internal strategic design reviews and is a frequent speaker at major international software development conferences. He is the author of several bestsellers, and his latest book is *Righting Software* (Addison-Wesley, 2019), which contains his groundbreaking ideas on system and project design.

Juval published numerous articles, regarding almost every aspect of modern software development and architecture. Microsoft recognized Juval as a Software Legend, as one of the world's top experts and industry leaders.

Acknowledgments

First and foremost, I'd like to thank my wife, Mirjam, for all her support over the years. This book would not have been possible without her. I am also very much indebted to everyone who has shared their knowledge with me such as teachers, colleagues, friends, family, authors, and countless others. Many people dedicate so much of their time to help others learn and progress; this is my chance to thank you all.

No one would have understood what I'm trying to say without the fantastic help of my main copy editor, Leila Johnston. Thank you so much for your skill, expertise, and attention to detail. You helped me with every aspect of this book and have been an absolute joy to work with.

Visually, the book came to life thanks to your beautiful illustrations, Danira Spahić—you understood exactly what I was looking for. I often outlined a rough idea in just a few words, and based on this you created illustrations even better than I could have imagined. Since finishing the final image, I've really missed working with you!

As a first-time author, I find it incredible how many people are prepared to spend their spare time helping with such a book project. A great big thank you to my mom, Carolyn, for always having emphasized the importance of languages. Your time and expertise in helping me realize this book are highly appreciated. My sincere gratitude goes out to Ben Smith for sharing his invaluable hands-on project experience. A massive thank you also to my dad, Tony, and my brother, James, as well as to Minh Vuong, Peter Smith, Charles Smith, Philipp Ochsner, Jon P. Smith, and Marc Mettler for proofreading the book. Thanks also to Ymyr Osman and Reto Scheiwiller for your wonderful ideas and inspirations.

Preface

Understanding what technology can and can't do has become a core competency that every part of the business must have.

—Gene Kim, Kevin Behr, and George Spafford
[Kim, Behr & Spafford 2013]

The meeting room had cleared when Daniel leaned forward and asked, "Johnny, what is a system interface exactly?" I was so glad he asked. The question lit a spark and inspired me to write this book. Daniel is a friend and former colleague who's well-educated in finance. He's the author of some of the best business requirements I've come across. We had cooperated on projects before, and I'm certain I had used the term "interface" a few times previously. Had I kept him in the dark all this time by overusing IT jargon? On a broader scale, how can we fill this knowledge gap between IT teams and business specialists? What resources are out there to help business people pick up the essentials? Not many, I discovered.

What, then, would be the best way to explain the basics of software projects? I began taking note of the examples I used for making technical topics clearer. After 8 years of collecting explanations, I've finally found the time to compile these into a book, and I'm delighted to now present my ideas to you. I aim to demystify the underlying concepts of software projects to let you in on the act.

Almost every company today is becoming a software company to some extent, yet software projects still suffer a high failure rate for a multitude of reasons. I'm convinced, though, that a better understanding between IT and business teams will help avoid common pitfalls.

As I've aimed to make this book an easy read, I must apologize now for any oversimplifications. The goal is to get the message across in simple terms rather than to cover every possible permutation.

In answering Daniel's preceding question, I used the metaphor of pipes and cables leading to a house. Instead of electricity or fresh water, a system will typically send data. I'll elaborate on this example later in the book.

Experience has taught me just how effective metaphors can be in illustrating many aspects of software. In discussions between techies and the business side, the right metaphor has the power to portray a complex technical topic in an instant. Examples based on the design and construction of buildings work particularly well. No one would dispute the idea. *If you want to add a new level to the Leaning Tower of Pisa, you'll need to secure the foundation first.* Likewise, before adding new visible features to a piece of software, you may first need to invest in the underlying code structure.

Introduction

What we do not understand we do not possess.

—Goethe

Helping teams achieve higher success rates in software projects is one of the main goals of this book. The key is to establish common ground on software project concepts among all the stakeholders in your project.

When creating software, some poor decisions are made that would never be made when building or renovating a house. Would you, for instance, create a beautiful new bathroom in a house that would be torn down in 18 months? No? I've worked on software projects where people made decisions just like that. If you were working on your own house project, you'd certainly take the time to check every aspect of it. Your business team really needs to do the same in a software project. Too often though, these teams lack the time necessary to collaborate well with the IT teams. Assigning them sufficient time to focus on the project will enable the business teams to think things through more carefully and make a fuller contribution to the project—optimizing the chances of project success. Also, while any serious company will keep its buildings in good condition, many tend to neglect some of their core business systems for years. By bringing the non-techies into the act, it follows that the mixed teams will make better decisions.

Metaphors, illustrations, and genuine examples can help reveal the core concepts. Some good metaphors can be found in techie books, yet I doubt any non-IT people would buy these and skip the technical stuff just to read the metaphor section. I therefore decided to write this guide specifically for the audience that would benefit most.

INTRODUCTION

I'd like to mention a book my new boss at a finance company sent me shortly before I started the new job. The book, on asset management, was incredibly helpful in giving me a grasp of many important topics and common terms. It sped up my learning quite dramatically when I started in the new role. Similarly, this guide is full of practical advice for people with little or no background knowledge in software projects. On the other hand, you may have participated in such projects already, but haven't felt quite at ease with all the technicalities. I've seen business people showing a keen interest when someone makes an effort to explain technical terms in an interesting way.

Having worked in the field for over 20 years, I've drawn on my experience and that of my extended network to present a collection of ideas that will be relevant to your software project, alongside working methods that I know to be effective. So far in my career, I've been lucky to work for some fantastic clients, from start-ups to blue chips. Although most of the software projects were implemented successfully, my focus here is mainly on the problematic ones. The examples here serve well to show you what went wrong and how you can prevent any similar issues in your own projects. As the book aims to cater for a wide audience, I find myself treading a fine line between describing things in plain language and trying not to alienate or bemuse the software community.

Finally, I'd like to highlight the importance of *communication*. Though relevant in all types of projects, this factor is, I believe, especially critical in software projects. How can teams make good decisions if not through effective communication and coherent terminology that's clear to everyone involved?

Who This Book Is For

Technology and digitalization is neither a threat nor an end in itself, but will provide us and our clients with added value at various levels. So, let's go for it together!

—Christoph Hartgens

Written with business people in mind, the book offers you a key to the world of software projects. An essential part of the digital transformation is about involving the business teams much more. The software projects need you because your decisions and involvement will be crucial to the project outcome. The book will not teach you how to program but will give you an overview of the steps and processes involved in creating a piece of software. Concrete, real-life examples will introduce you to the basic concepts, with a focus on your role and your deliverables at the same time. Whether you're a subject matter expert, a manager, or a user representative, you'll find this guide invaluable. Your newly acquired knowledge will help you reach the market faster and meet your customers' needs far more effectively.

While researching for this volume, I was astonished to find that very few books on software address this audience. That's why I aim to give you a complete picture of how the various parts of a project fit together.

 Has your boss nominated you to represent your department in a software project? Do you need to review and sign off formal IT project documents? Have you felt overwhelmed in meetings when an IT specialist reels off a stream of IT jargon?

The information here will also enable project sponsors and line managers to gain a better insight into project best practices. The rich set of illustrated themes will help visualize the common steps. Even beginner software developers can enhance their knowledge of the more practical side of projects. A project manager, meanwhile, could present copies of this book as essential reading to the team.

INTRODUCTION

Company executives know they need to understand both the potential and the pitfalls of IT. A company may need to reshape its digital strategy, for example. This digital transformation must be carried by everyone in the company. One of the most important developments is in bringing traditional businesses into the digital age. IT cannot do this alone—the business teams must also be involved. At the same time, business-critical software projects will require some executive decisions. Therefore, specific sections focus on helping executives make decisions that are well-informed.

The technical project members involved in a software project are often outnumbered by the business professionals—and their input is a great asset in any project. To fully utilize this business knowledge, the team must know how to apply it best. Team discussions and decisions will be very effective when all members understand the core concepts of both the business and technology sides. One of the biggest barriers that people face is, in fact, the techie language, which is mostly incomprehensible to anyone without a background in computing. You may feel too embarrassed to comment or ask questions in the face of it, even if you have a valid contribution to make.

Learning the basics of software projects is therefore a bit like learning a new language. The more you understand, the more involved you can be. Similarly, by broadening your software project know-how, you'll be able to participate effectively—who knows, you might even start to enjoy the projects (more)!

Business and IT teams that communicate well together are incredibly powerful—this essential element of good communication is often the missing link in unsuccessful projects. Also, by applying best practices, you'll enable your business to adapt better to changes and keep its competitive edge. These days, we witness how traditional businesses such as book stores, taxi services, and record companies are shaken by global software solutions. Which sector will be next? All businesses really need to be software savvy now to survive.

I sincerely hope this book will help you build a strong foundation for your software projects. Lastly, I welcome your feedback and would highly appreciate your participation in this exciting topic. Please visit the book website at www.SoftwareGuide.blog.

How to Use This Book

In using a conversational tone, I imagine that I'm interacting with you directly. My aim is to make the information as accessible as possible— enabling you to rapidly increase your knowledge and engage effectively in your next project.

Structure of the Book

The book is divided into three main parts. Part 1, "Conceptual Guide," will help you understand the main concepts behind software development. The metaphors offer a high level of abstraction and allow you to understand something new much faster. Not only will you feel more at home in an unfamiliar place, but you'll also gain a more holistic perspective on software projects. I make comparisons between software development and physical construction projects, as everyone can relate to building a house. After identifying the commonalities between renovating and building from scratch, we'll look at the differences between the two. In software as in physical construction, creating something new can vary distinctly from modifying a structure that already exists. Finally, we'll explore some further metaphors as we extend beyond the construction comparison.

Part 2, "Practical Guide," focuses on best practices in the hands-on side of software projects, both large and small. We'll look at the whys and then the hows. This part of the book, which runs through the various stages of a project, focuses on the topics most relevant to you, the business professional. A brief outline of the technical side of things will give you a bird's-eye view of what goes on behind the scenes at the same time.

Part 3, "Technical Guide," digs a little deeper into some of the common technical topics that all projects need to address. The relevance of this part of the book to you will depend on your role and interest. Again, metaphors and analogies help describe the technical concepts in a clear and interesting way.

I introduce the necessary terminology gradually, giving you the chance to become familiar with the terms. "Appendix B: Glossary" at the end of the book explains the technical terms in plain English. Additionally, the appendixes expand on some of the topics introduced in the main part of the book.

References

The quotations and extracts, all referenced, are based on best practices or on research findings. Also, when you find a topic of particular interest, you can delve into some of the related materials listed in "Appendix C: References and Further Reading." There, you'll find a selection of books, online resources, and videos to choose from.

The references to source materials are presented as: [Horowitz 2014], for example. Additionally, if an author has asked me to include the page numbers, then this format is used: [Brooks 1995 p. 55].

The Value of Metaphors

The value of metaphors should not be underestimated. Metaphors have the virtue of an expected behavior that is understood by all. Unnecessary communication and misunderstandings are reduced. Learning and education are quicker. In effect, metaphors are a way of internalizing and abstracting concepts, allowing one's thinking to be on a higher plane and low-level mistakes can be avoided.

—Fernando J. Corbató

Before digging deeply into any unknown topic, it's important to build a mental framework that helps put things in place. The powerful effect of metaphors and analogies makes them indispensable when explaining software projects. Here, we'll take a brief look at the general topic of metaphors.

As a leadership coach and author, Dr. Peter Fuda states

- "metaphors stimulate creative thinking by inviting the reader to discover complementary and related meanings and applications

- metaphors make complex stuff simple by introducing you to an idea and making it much easier to explore once you're inside it

- metaphors use familiar imagery and hence make a topic easier to recall"

[Fuda 2012]

Now, without further ado, let's dive in and begin with a look at the main metaphor used here—**a software project is like a construction project**.

PART I

Conceptual Guide

CHAPTER 1

Architecture and Construction

In Part 1, "Conceptual Guide," we look at the tasks involved in building software that are similar to those in physical construction. Software architecture is named thus for a reason—you'll see many similarities between designing software and physical buildings. The architect's team will create plans for all the construction work including things like the electrical wiring, the plumbing system, the landscape design, and so on. The same goes for software. In this case, the architect considers data flows, user interaction, sequence diagrams, and many other elements in the process of planning, discussing, and building. You'll find more information on the various roles in software projects in "Appendix A: Collaboration."

What happens if you radically change your mind about the house you're having built during the planning phase? That's doable. Parts of the building will have to be redesigned, and costs will need to be adjusted; but it's certainly achievable. How about if you change your mind when the building is half completed? That's a lot trickier to deal with. At best, the changes can be artfully worked into the ongoing construction; but in the worst case, most of the building might have to be torn down and rebuilt. That's not to say it can't be done—just that it'll be very costly and time-consuming. Although modern software projects cater better for late changes, radical ones will continue to cause higher costs and delays.

© Jonathan Peter Crosby 2023
J. P. Crosby, *The Business Manager's Guide to Software Projects*,
https://doi.org/10.1007/978-1-4842-9231-0_1

Treating software construction as similar to building construction suggests that careful preparation is needed and illuminates the difference between large and small projects.

—[McConnell 2004]

As the home buyer, you may not care much about the exact route of the drainage pipes in your new house, but you'll be very interested in the layout of the kitchen. The construction plans will be adjusted to reflect your choices. Similarly, some areas and details in software design are more relevant to the non-IT professionals than others—you probably won't need to know about all the "plumbing" or the "under the hood" techie stuff.

In construction, the interconnection of the various building elements and materials is fundamental. The materials must be weatherproof and also easily replaceable. The paint, windows, wood, and bricks all have different lifespans, for instance. The windows may need replacing after 25 years, but the bricks will typically last much longer. Software parts have different lifespans too. The support for a software component may end next year and will therefore need to be replaced, for example. In both areas of work, the architect and development team need to make sure the structure will be sufficiently functional to keep all the parts independent and updateable.

Additionally, just as a residential area isn't designed to be converted into a theme park later, software cannot easily be converted into something much bigger either. The architect needs to start from a new plan in both cases.

Inevitably, the fiddly bits end up taking the most time. Any professional floor tiler will confirm how the shaping of the small pieces takes a lot longer than laying out the whole tiles. The trimming work is also a big part of software development.

Despite all the planning and preparation, unexpected factors will almost certainly crop up and create extra work. In a rather extreme case, the house of a former colleague of mine began to sink while still under construction. The architect arranged for 12 large concrete piles to be driven into the ground around the house to stabilize the foundation. Another kind of complication would be the discovery that the building you were about to renovate was a heritage building. The structure would be subject to certain laws, and you'd probably need to comply with additional rules, adding further costs. You might be inclined to think that nothing comparable could happen in software projects, but you'd be surprised. Many unforeseen factors can suddenly appear and necessitate extra work. The team may have overlooked an important stakeholder or a critical business case, new regulations may have come into force, or a security breach could be discovered, to name just a few.

Note   To ensure that the terms used throughout the book are clear, I will use the word *construction* when referring to physical construction and *development* when describing software development.

As in most human activities of any complexity, a software project begins with planning...

CHAPTER 2

Planning and Scope

All projects require some degree of planning, depending on the size and complexity of the project. At its core, a project is about managing an abundance of small decisions and dependencies. This requires high-quality communication on all levels. And just as a physical model can help us visualize a new building, creating a working prototype is also useful in software development. This prototype gives users and stakeholders a good idea of how the product will look. Modern software methodologies usually create screen mockups rather than prototypes—these mockups are easier to produce and give a good visual impression.

The planners, at the same time, will need to conduct a survey before embarking on a project plan. Just as a feasibility analysis should first be carried out to check the viability of digging a tunnel, the same needs to be done for a large software project.

> *Feasibility studies are preliminary studies undertaken in the very early stage of a project. They tend to be carried out when a project is large or complex, or where there is some doubt or controversy regarding the proposed development.*
>
> —[Feasibility Studies 2017]

© Jonathan Peter Crosby 2023
J. P. Crosby, *The Business Manager's Guide to Software Projects*,
https://doi.org/10.1007/978-1-4842-9231-0_2

The Right Dimensions

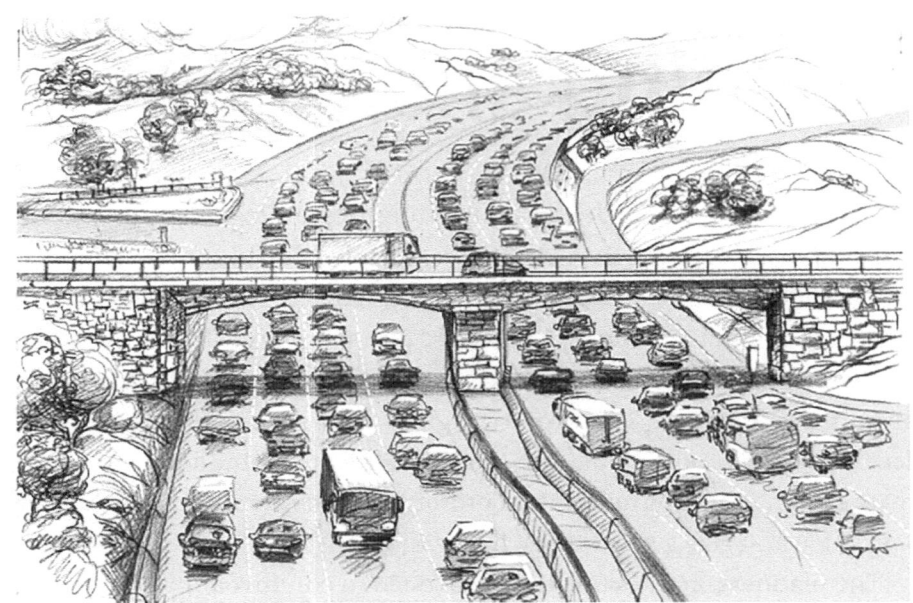

Figure 2-1. *How many lanes should you plan for?*

When building a new highway (Figure 2-1), one of the things the planners will need to decide on is the width of the bridges that will cross it. Let's say that the highway will initially comprise three-lane roads in both directions. Should the planners *future-proof* all the bridges to leave scope for adding another three lanes later? How would these wider bridges affect costs?

If the roads can currently cope with around 2,000 vehicles per hour but the number increases to 10,000 after a few years, then further construction may be required. This situation is comparable to the number of concurrent users of a software application. Software for creating team reports would naturally require a leaner technical setup than an online ticket system where 200,000 people may be trying to buy concert tickets at the same time, for instance.

Although future-proofing might seem a good idea, it can sometimes backfire. Here's one example: Some 30 years ago, a highway tunnel was built under the main railway station in Zurich, Switzerland, in anticipation of a future road. Today, there is still no road connected to this segment, and there probably never will be. Imagine the complexity of building a tunnel underneath a large station, as well as the high cost incurred. Trying to foresee the distant future is equally hard in software projects. Making absolutely everything configurable to allow for future changes is not always the best approach—it will make the software a lot more expensive to produce and maintain.

When building a new house, the homeowner may request the builder to place empty tubes in the walls that will be useful later if additional cables need to be laid. If not required, these tubes can usually just remain in place with no extra work or cost involved. The opposite is true for software, however. Because every new feature needs to be tested and maintained, features that aren't needed now should not be built.

Hogwarts Castle—Keeping Within a Budget

Some people may dream of living in a castle or a palace (Figure 2-2), but common sense usually dictates that you build something more modest with an affordable budget. If you plan to build your own house, you'll naturally think very carefully about what you need and what you can afford. The initial outline will include the size and type of house you want and the number of bedrooms and bathrooms you require. This process will involve some hard decisions. Can the kids share a bedroom? Do I really need an office? Can I afford a pool? How much can we spend on the kitchen? You'll also consider hundreds of minor details. Do I need a power socket here or a light switch there? Your house project will include an abundance of decisions both large and small. Some decisions will be hard to make, and not every wish can be fulfilled.

Figure 2-2. *Choose the right size for your project*

The same is true for software projects, but with one major difference—in most cases the company pays the bill. Some project members try to build a Hogwarts Castle with the budget for an igloo. Seriously though, in larger corporations especially, a project member may have a vested interest in building a beautiful castle, or they just struggle to adjust the scope despite budget restrictions. The person may pursue this idea even when having software with all the bells and whistles is not in the company's best interests. Across the projects I've worked on, there have often been business colleagues who treated every request as a must-have. This habit makes tough calls even tougher when budget and time restrictions eventually force the teams to make cuts to the project.

Note   A countermeasure to this common issue of scope reduction is to focus on building the minimum viable product (MVP) from the start.

This minimalistic approach basically involves developing only the software features that are crucial for the upcoming version. Even then, it's hard work deciding on which features and tasks will be done in what order. These decisions require continuous collaboration with the customers, the stakeholders, and the development team(s). This prioritizing of work is one of the most critical tasks of the product manager. Just putting all the customer's wishes at the top of the priority list simply won't cut it.

The way in which your teams will work largely depends on the type of methodology chosen for a particular project. The next section will show you how to approach the decisions involved.

Working Together

Having efficient teams is essential in any type of project. As software development is still a young industry, especially when compared with construction, identifying the best way for teams to work together is still a work in progress. A large software project can involve many stakeholders from different departments—together they need to make an abundance of decisions. Keeping track of these decisions and ensuring that all the relevant team members are informed does pose quite a challenge. Let us look at some of the most common scenarios.

Ski Resort—Software Development Methodologies

Imagine you were responsible for planning, building, and running a new ski resort (Figure 2-3). You could build all the lifts and slopes for the whole resort at once and open the place with a big bang. Alternatively, you could open the resort with a limited but fully functional set of lifts and slopes. This second option would allow you to open earlier and then gradually grow to a larger size. Both approaches come with certain advantages and disadvantages. For the first model, you'd naturally need a larger upfront investment before you could gain any revenue or customer feedback, whereas the initial lack of variety in the second model may not attract a wide base of customers.

Figure 2-3. *Gradually grow or build big from the start?*

These two approaches can be loosely compared to the *Waterfall* and *Agile* software development methodologies. The traditional Waterfall method is also named the *big bang approach* as it aims to deliver everything in one go. Accordingly, the project is done in sequential steps that comprise analysis, design, construction, testing, production, and maintenance. The Waterfall approach is named thus because it flows steadily on from one step to the next. The Agile approach, on the other hand, uses shorter timeframes called *iterations*. Each iteration is itself like a mini-project that includes all the steps from analysis to a fully functional piece of software. It is then for the business department to decide if the outcome of the mini-project warrants a new software release into production.

Most project frameworks fall roughly into one of these two methodologies. Although there are many variations of these two project structures, for the purposes of this book, I'll focus on these two most commonly used work methods. We'll take a closer look at these in the "Methodology" section of Chapter 12.

Cranes—Knowledge Workers

Figure 2-4. *Minimize the task switching*

The scene: A building site with three cranes and one crane driver (Figure 2-4). The driver climbs into one of the cranes, does 5 minutes' work, climbs down, and then gets into another crane. They work for another 5 minutes before going on to repeat the process in the third crane. After that, they return to the first one. And so on. Every switchover takes 10 minutes. Meanwhile, other workers on the ground await the delivery of materials. If the supervisor saw this constant rotation and loss of productive time, they would certainly look for a better method. Couldn't the work at least be organized so that the crane driver could work for longer stretches in each crane?

A similar situation can emerge in software development.

Programmers are regularly distracted and are frequently expected to switch tasks. For cognitively demanding tasks, the time it takes to resume the original task can often be considerable.

Clearly, all knowledge workers need to maintain a high level of concentration when performing the main tasks of their professions. The work of managers and project leaders is often the opposite, however. Their role requires flexibility while overseeing others' work, participating in meetings, taking calls, and so forth. Managers therefore—especially those who've never worked as programmers—may not understand how much time is lost when programmers are constantly having to switch tasks.

Note   Imagine you're counting a huge pile of coins and keep getting interrupted before you have a chance to note down the current figure. Wouldn't you ask for a few minutes to finish counting?

CHAPTER 3

Teething Troubles

To round off this portion focusing on commonalities between creating the new and changing the preexisting, we look at the handover to the buyers and what can go wrong.

Every house and every piece of software will have some teething troubles after completion. Once the house keys have been handed over, the furniture has been delivered, and daily life begins inside, the occupants of the property are bound to discover some problems. You may find scratches on the new floor, improperly wired wall sockets, or a door that sticks. The process of settling in and organizing everything to your satisfaction is often gradual and organic.

If any serious problems with the new house crop up, the building company will need to fix the issues as part of their guarantee. If, on the other hand, the problem is due to a choice made by the property owners, the guarantee will not cover this work. Likewise, if software doesn't work as it was originally ordered to do, such problems are classified as bugs. Other modifications will result in a *change request*. Differentiating between a bug and a change request can determine who must pay. It's very helpful to have well-specified requirements in this situation—this information will keep the discussions about these common problems factual. Remember that tracking these issues—and having sufficient time and resources reserved to fix them—is a critical part of any project. The level of support after the *go-live* and the level of ongoing maintenance required will usually be specified in a contract.

© Jonathan Peter Crosby 2023
J. P. Crosby, *The Business Manager's Guide to Software Projects*,
https://doi.org/10.1007/978-1-4842-9231-0_3

Even beyond these initial teething problems, however, the software still cannot be regarded as completed. The development team may be needed further down the line because various issues could still appear. There could be a decline in the software performance as the number of users and the amount of data increase, for example, or new security threats or bugs could cause the system to fail completely. The maintenance team may not have the time or skills to change the software code.

I've personally experienced such problems arising when software was handed over to a maintenance team shortly after the go-live. You can imagine what a headache this can be when no developers with knowledge of the system remain. This situation occurs when companies think and plan in terms of completed projects rather than ongoing products. The management has not, perhaps, considered how time-consuming it is for someone new to get to know the software. The newcomer will need a lot of time to understand the software design before making the changes necessary to fulfill the quality expectations. Without a proper handover from the existing development team to the new developer, the learning curve is much steeper. Ensuring that the development know-how remains available after going live is therefore critical, and it needs to be well-documented.

As prevention is always better than cure, I'd next like to show you the type of mistakes made at the outset that can cause big disruptions later in the project and how to avoid them.

Conceptual Mistakes

The most expensive mistakes are the conceptual ones made at the start of a project. The core decisions made, then, will affect a large part of the work to be done. If the team discovers any mistakes early in the development process, they can prevent any damage. When mistakes are not detected until much further down the line, however, you may even need to restart the entire project from scratch. Unresolved conceptual issues can result in a faulty system. Short-sighted and temporary fixes typically result in high maintenance costs, not to mention significant compromises that affect the users. Let's consider three real-world examples of costly conceptual mistakes before looking at the measures you can take to avoid such mistakes.

Conceptual Mistake Example 1— Different Measurements

Figure 3-1. *Measure twice. Cut once*

The case of the Hochrhein Bridge well illustrates the results of a mistake made at the start of a project (Figure 3-1). Built in 2003–2004, the bridge crosses the river Rhine, linking Germany and Switzerland. Both countries measured in meters above sea level. Germany used the Amsterdam level, and Switzerland used the level of a rock in Geneva. This rock, in turn, is based on the sea level in Marseille, France. A difference of 10 1/2 inches existed between the two measurements. Although the engineers were aware of this difference, they made an error when compensating for the difference–reducing the height of the bridge on the Swiss side by 10 1/2 inches instead of increasing it. This miscalculation made a total disparity of 21 inches in height between the two sections. When the mistake was discovered during construction, the engineers applied a gigantic fix on the German side of the bridge, and the structure was completed. Luckily for the engineering company, their insurance covered the resulting costs.

Conceptual Mistake Example 2—Where's the Restroom, Please?

Even experienced architects sometimes make mistakes and forget things. Those attending the 1991 opening ceremony of a newly built railway station were in for a big surprise. There were no restrooms. This essential facility at the new Kassel-Wilhelmshöhe station [Kassel–Wikivoyage] in Germany had simply been forgotten. The restrooms were added later, of course, but took up part of the area meant for a bike store, which ended up much smaller than planned as a result.

This kind of suboptimal workaround can permanently harm the revenue of a business. Also, the cost of fixing a conceptual mistake must be weighed against the cost of restarting part of the construction.

Conceptual Mistake Example 3—Where's My Suitcase?

You may remember hearing about the debacle at London's Heathrow Airport when software design issues in the luggage handling system made global news. The software had sailed through all the preliminary tests at the newly opened Terminal 5. Then on the first day of opening, the whole system crashed. It turned out that the testing hadn't covered *all* real-life scenarios. The system couldn't process the data when the luggage needed removing if someone didn't board the plane. This simple action confused the entire system and caused it to shut down. Consequently, over 40,000 pieces of luggage were left behind after flight departure, and some 500 flights had to be canceled over the following days.

Conceptual Remedies

Keep in mind that mistakes *will* happen when people work on any type of project. You need to look at how to minimize them. You'll find that careful planning and execution, together with clear and open communication, will help reduce and resolve mistakes. Always do your utmost to identify mistakes as quickly as possible. Even with every precaution though, big and expensive mistakes will occasionally occur. Then, it's important for all team members to stay level-headed and look for solutions together. Most failed projects are in fact due to sociological rather than technical reasons. Finding a solution instead of blaming each other is in the best interests of the project and the company.

Many major software architecture defects are never actually published for fear of reputational damage to the organizations involved. Errors that affect the public are reported much more often. Fortunately, there is a growing trend for companies to be more open about their internal mistakes. Everyone will benefit when employees also feel they can report their own mistakes without fearing a name, shame, and blame reaction. We'll take a closer look at this topic in the section "Fail Fast" of Chapter 12.

Fixing Issues Centrally—Why Is the Tap Water Dirty?

If dirty water is flowing from every tap of every apartment in a building, the root cause will come from a shared component. Once identified, the problem can be resolved at its source. A pipe leading to the building may have been damaged by nearby roadworks, for instance, or some of the common pipes may be rusty. In this case, requesting every apartment owner to install a water filter system would hardly be a viable solution. Strangely enough though, I've seen the latter logic used in handling certain issues in software systems.

In one scenario, a central system delivered the same data to various recipient applications. Each receiving application was responsible for its own data quality and checks. By using different data guards across different systems and automated corrections of false data, the data diverged. This divergence can lead to substantial extra work and new issues, especially if the data is combined further down the line for reporting. A much better solution is to resolve data issues centrally, avoiding any replications of work by finding a single solution from which all data recipients can benefit.

Beauty vs. Practicality

Beauty and function are by no means mutually exclusive—we all love beautifully designed things, especially when they have a practical use too.

Note   Beauty and function are by no means mutually exclusive.

Sometimes functional shortcomings are not discovered until a while after the handover to the buyers. The faulty bridge in Italy described in the following is a case in point.

In Venice, a city with 409 bridges crossing a total of 177 waterways, the place should know a thing or two about building bridges. The new bridge *il Ponte della Costituzione* looks spectacular—I personally like the contrast between old and new. The structure is, however, a classic case of form over function. The bridge has proven to be very dangerous in wet or icy conditions—the glass paving has caused many accidents. Would the architect themself like to cross this bridge every morning to get to work?

In a different case, a friend of mine who works as a nurse was telling me how their new hospital building looked great from the outside but was impractical as a place to work. The furniture and equipment did not fit into the rooms well, resulting in wasted space and areas hard to access. By way of contrast, a recent TV documentary showed a new hospital in the Netherlands that was designed around the daily processes involved. The routines of laundering the bed linen and hospital clothes were mostly automated, for example, and this process optimization achieved large cost savings.

The same applies to software. A design that is fabulous to look at but hard to use will be purely decorative—its core purpose will simply not be met. On the other hand, if the software looks awful but is highly practical, it can still be used for the intended purpose. Overemphasizing the design while failing to achieve the purpose of the software will leave the users feeling frustrated.

As a user and supporter of applications that have suffered from poor architecture decisions, I've often wished those architects had to walk a mile in my shoes. It's no coincidence that the best software architects I've worked with have spent some time supporting systems. In other words, they had gained hands-on experience on the practical side of software systems.

Returning briefly to the example of the badly designed hospital, the architect would benefit by spending a few days observing some of the hospital staff at work. Understanding the tasks and working methods of the nurses, doctors, and cleaners would give the architect some valuable insights before finalizing the design. Such preparation would enable the architect to address function more specifically and effectively.

Taking this idea a step further, Christopher Alexander describes a rather profound concept in his book *A Pattern Language*:

> *At the core of this book is the idea that people should design for themselves their own houses, streets, and communities. This idea may be radical (it implies a radical transformation of the architectural profession) but it comes simply from the observation that most of the wonderful places of the world were not made by architects but by the people.*
>
> —[Alexander et al. 1977]

I've had the pleasure of working with some great software architects. However, in some cases there was a considerable gap between their theoretical view of the system and how it actually worked in production. Also, another gap often exists between the software developers and the maintenance team. One of my clients began rotating the support team with the developers to better address the software issues in the live system. The results were astounding—the developers improved their code and suggested new ways of reducing the support effort. The founder of the tech company Dynatrace went one step further and made the architects fully responsible for the production system support. They designed better software as a result, not wanting to spend their time fixing production issues.

In the next chapter, you'll get an idea of working on greenfield projects, where there is no legacy system to worry about.

CHAPTER 4

Greenfield Construction

Figure 4-1. *Building anew*

© Jonathan Peter Crosby 2023
J. P. Crosby, *The Business Manager's Guide to Software Projects*,
https://doi.org/10.1007/978-1-4842-9231-0_4

Building a house in a field with no adjacent buildings—where cranes and materials can easily be brought in—is quite different from renovating a house in a city street (Figure 4-1). For the latter, existing walkways may need to be secured, and the delivery of materials will be much harder. In software projects, having no preexisting dependencies to worry about is like working in an open field. It will considerably reduce the effort and duration of the project. The term *greenfield project* is widely used to describe new projects, including software, that are free from any constraints caused by prior work.

Building a small tree house for the kids is really getting down to basics (Figure 4-2). As you'd expect, it would require little planning, and any changes would be easy to make. The materials for the tree house could be purchased and assembled according to a simple drawing and just a few measurements. Building a house or bigger building, on the other hand, would naturally require a lot more planning and work. At the risk of stating the obvious, these different types of structure and degrees of complexity also apply to software projects both large and small. Unsurprisingly, perhaps, research shows that small software projects have a much higher success rate than large ones.

Most educational institutions teach students how to build new software from scratch. This theoretical knowledge is useful to have; however, in the real world, most juniors will start by extending existing software. A lot of software is of poor quality or doesn't adhere to best practices. The programming courses offered by universities and other studies should cater more for this situation, ideally also showing a path to take poor software and improve on the quality.

Figure 4-2. *Building a simple tree house requires much less upfront planning than building a real house*

Now, we'll turn to the question of whether you should buy or build the software for your project.

Houses and Cable Cars—To Buy or to Build?

When you plan to build a new house, you'll need to decide whether to buy an off-plan house or have one built according to your own custom design. The deciding factor is quite simple: does the off-plan solution fulfill your needs? House buyers who plan to run a home business may require special layouts for the rooms. A music teacher might want a student area with its own entrance that's separate from the private family area. It will be harder to find off-plan houses that meet such specific needs.

The same is true for software. You need to ask, "Does the software need to be custom built for our business?" Somewhat surprisingly, I often see the wrong decisions being made here. This question should be evaluated with the development team and software architects once your core requirements are clear. Generally speaking, the choice depends on whether your company has industry-standard business processes in place. The off-the-shelf software on the market will typically cover most standard processes.

Note   The more unique and specialized the requirements, the more likely it is that only a custom-built solution can fulfill your needs.

In some cases, your company may want to evaluate whether they can standardize their business processes before choosing any software. Customizing a standard piece of software too much is self-defeating. It will become awkward to use and end up being costlier than a custom solution would have been, but this is a common mistake. This approach also carries the risk of losing the major benefit of packaged software—in easily upgrading it to a newer version.

To understand the customization process involved in packaged software, imagine a cable car being built to transport people to the top of a mountain (Figure 4-3). The cabins, cables, and various other parts may be standard, but the pillars need adapting to the terrain. Similarly, large software toolkits may have a standard core and adapters for connecting to preexisting systems. Customized connectors can adapt to other systems and workflows to represent specific company processes. Of course, the more customization you require, the more time, resources, and money you'll need.

Which option to choose is an important decision that needs to be made for every software project. Therefore, I've elaborated further on this topic in the section "Buy vs. Build" of Chapter 13 in Part 2, "Practical Guide."

As the focus of this book is on projects that build custom software, we can now have a closer look at the various aspects of building.

Figure 4-3. *Some parts are standardized—others are custom built*

The next section will focus on the mockups—these give a good idea of how the visual part of the software will look.

Hollywood Façades—The Work Behind the Scenes

Hollywood is known for creating illusions. The houses, towns, and even cities you see in movies may well be mere façades constructed in the studios, but the level of detail involved can appear very impressive. Only the parts that will be visible to the cameras are built, and one look behind the façades will reveal the empty spaces. These façades fully serve their purpose, however, and save a lot of time and money.

Visual mockups, which act in a similar way to façades, are often created early in a software project. These give the business department an idea of how the user interface (UI) will look. The mockups can look just like the real thing and serve to elicit important early feedback from the business. Although a mockup provides no functionality, I've known clients who've greatly underestimated the amount of work still to be done once the visual elements were decided upon. The functionality work behind the façade actually takes a long time to build, as the following metaphor illustrates (Figure 4-4).

Figure 4-4. *Building only a façade is much faster*

When a skyscraper is being constructed, you may have noticed how quickly the tower is built upward. The façade and the windows can also be mounted pretty fast. By contrast, the initial foundation work and the interior work done later take a lot more time. The many different steps of this interior work will begin once the bare construction is completed. Dozens of different specialists will contribute to the more detailed and time-consuming work.

Figure 4-5. *The bulk of the work is hidden*

Similarly, coding the functionality will usually take much longer than creating the visual design elements. Although every profession involves a lot of work behind the scenes, programming seems to entail more hidden work than most. No real results can be seen until the code is running with a visible result. In fact, you can imagine a software project as something like an iceberg—the bulk of the iceberg that's hidden underwater represents the programming work that is unseen (Figure 4-5). The tip of the iceberg above the surface is like the visible part of software that you can use and hence relate to more easily.

CHAPTER 5

Laying the Right Foundation

Having built the foundation for a family house, you can't suddenly change your mind and build a skyscraper on the same foundation instead. Getting the foundation right in software is equally important. It all depends on making the right conceptual decisions first.

This truism was made very clear to me early in my career in the following scenario: "Jonathan, our team really likes this piece of software. Could you please make it available to the whole region?" This challenge came from a team head during the second part of my internship at Barclays Bank. Originally, the assignment entailed enabling a small team to collect and track facility management issues. Coincidentally, I had just spent the first 6 months of my internship at the IT help desk. Understanding how issues were tracked had given me hands-on experience that was useful in completing this task. I had learned which data needed to be entered during a support call and how the software would ideally track an entire support case from start to finish. My experience in asking the right questions about the scope was still rather limited though. I might have designed a useful tool, but hadn't questioned their definition of the scope. I had understood that the software would be used for a small team in one location using a shared team network drive. Hence, I developed a small database (DB) with a user interface that worked on their network drive. When I presented the result, they were so

© Jonathan Peter Crosby 2023
J. P. Crosby, *The Business Manager's Guide to Software Projects*,
https://doi.org/10.1007/978-1-4842-9231-0_5

pleased that they wanted to make the software available to multiple teams immediately. They couldn't quite understand my technical explanation of why this wider requirement would involve a lot more work. Despite their compliments and their wish to use the software on a larger scale, I felt I had missed the target as I hadn't got the scope right.

If presented with the same challenge today, I'd use the analogy of the skyscraper foundation given previously to explain the scenario to them. Such a substantial change in scope would mean I'd need to build an entirely new foundation using a different product to handle multiple locations and more people. Finally, we'd need a role concept as well, to define which teams would support which office locations.

This example shows how a change of scope can dramatically affect the entire project. But without clear explanations, some non-techies may struggle to fathom the greater effort necessary.

Meanwhile, does your company like to play it safe or to experiment with something new? The following section will give you an idea of what an experimental project can involve.

Sydney Opera House—Experimental Projects

From time to time, some maverick steps into the limelight with a startling and innovative new design. The Sydney Opera House is an excellent example (Figure 5-1). This iconic landmark is one of the first images that spring to mind whenever Sydney or even Australia is mentioned. The fascinating history of this unique building is full of lessons for architects, construction engineers, and also software developers. I can highly recommend the Discovery Channel documentary on the history of this building in the *Man Made Marvels* series [Wonders of the World 2015]. There are so many parallels between this construction project and large software projects on both the technical and sociological levels. Here are a few outtakes.

Figure 5-1. *Are you planning to experiment with something completely new?*

Even with today's technology, constructing the so-called "sails" on the roof of the Opera House would be a challenge. When the design first appeared in the 1950s though, the task bordered on the impossible. Finding a shape that would prevent the sails from collapsing proved very challenging. Over a period of 3 years, Danish architect Jørn Utzon used every trick in the book as he tried and tested the project. Following the initial failures, he planned more elaborate structural tests in which the shape would allow weight and force to be distributed evenly across the building. The task was so big and complex that the earliest computers were brought into use for the first time in a construction project to crunch the numbers. The story goes that the peeling of an orange had led to the inspiration for shaping the sails. As one of the structural engineers for the

Sydney Opera House, John Nutt states in the documentary, "Anybody can make a complicated thing look complicated. The clever thing is to make a complicated thing look simple."

Utzon and his team had to develop many new techniques specifically for this building project. One such challenge was to devise a construction method for creating the sails that allowed both precision and mass repetition. Once the new "continuous casting" technique was developed, they produced over 2,000 segments. The Sydney Opera House then became one of the first large-scale construction sites to use such a technique for molding steel into uniform shapes. When the cranes proved unequal to the task of mounting the segments, a unique curved arch was designed to facilitate work at different angles and heights. As an entirely new process, this mounting work was anything but easy. Initially it took 3 days to add one segment, until eventually 14 segments could be mounted in a single day. As the narrator of the documentary says, "The Sydney Opera House isn't just a marvel of design; it is a marvel of construction as well."

Utzon's plans for the Opera House interior also involved complex shapes, materials, and procedures, all requiring extensive research and planning to develop. In breaking entirely new ground in the world of design, his quest for perfection naturally involved quite some time and expense. Meanwhile, the state government of New South Wales, as the body financing the project, became unwilling to continue their support. Utzon resigned and left Australia. When supporters held protests to bring him back, the architect agreed, but the government stood by their decision. It was subsequently a matter of urgency for the newly assigned architects to understand every detail and concept behind this, one of the most complex buildings ever designed. Controversy and criticism followed their every move.

"If Matisse had left a painting incomplete, who would ever finish it? Picasso wouldn't have dared."

Scrapping all the interior work that had been done, the new architects essentially destroyed all the harmony of the building. Although construction work on the building was officially completed in 1973, for many the masterpiece is still unfinished today.

In 1999, a delegate from a Sydney-based architecture firm went to Denmark to consult Utzon. The firm wanted him to set out his original principles for the design to allow anyone working on the building in the future to use that design as their guide. Despite his poor health, Utzon, with his son's help, succeeded in setting out the design principles for the Opera House for posterity. In 2007, the building was awarded World Heritage status, making Jørn Utzon the only living architect to have ever received this honor.

Moving away from this monumental building, we can say that such difficulties apply to some extent to all first movers and innovators. The same also applies to the early adopters of anything that is new and radically different. A company or project may benefit hugely from offering or using something new. It stands to lose a great deal though if the work ends in failure for any reason. At the same time, the more experimentation and innovation entailed, the more time and costs involved.

Will your project include any experimental elements, or will you use established paths and tools? Do you have the budget, time, and patience necessary to build the software equivalent of the Sydney Opera House? Would you allow one person to hold all the knowledge of the project and hence the key to its success?

Pushing back the limits by using the latest innovations can be very exciting. But utilizing new methods in software development can cause painful teething troubles too. Also, simple maintenance tasks could require individual solutions, incurring higher costs initially as well as for the ongoing maintenance necessary.

Even modern experimental building projects like the Elbphilharmonie in Hamburg, Germany, can incur cost and schedule overruns (Figure 5-2). Yet it was well worth the wait—the city's new blockbuster concert hall is extremely popular. People from all over the world flock to the performances and to view this spectacular building.

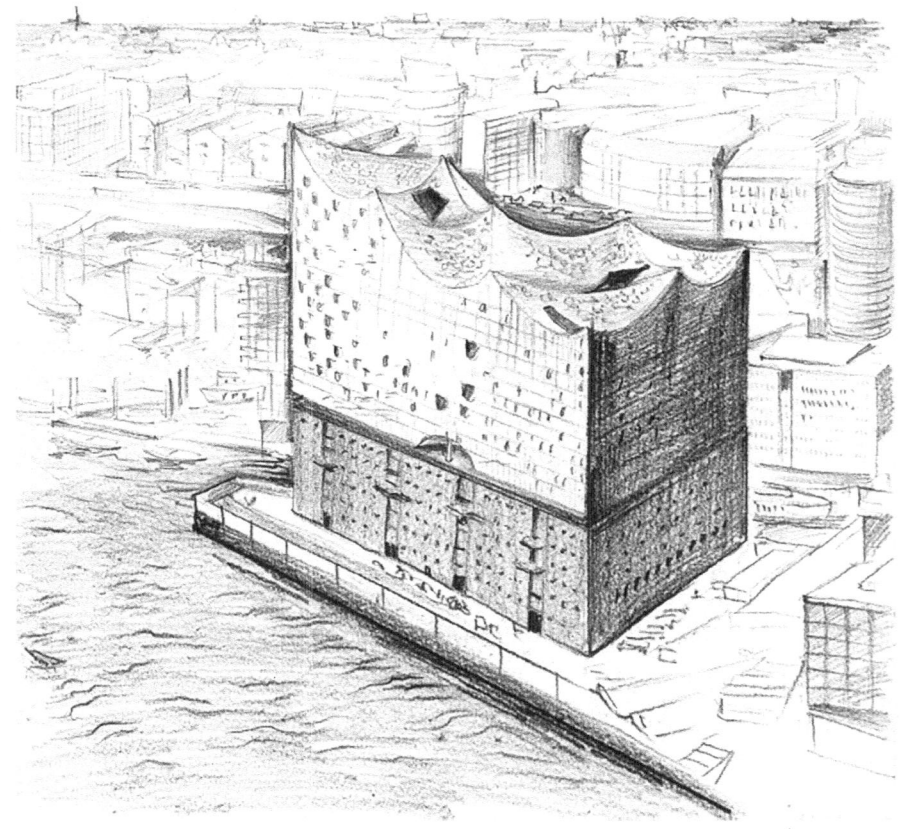

Figure 5-2. *Spectacular new designs are prone to delays*

To summarize, it's true that new technologies can address existing issues, but it's important to consider whether these new technologies lead to new challenges. If you do venture off the beaten track, you'll need to adapt your project plan accordingly. Using tested and stable standard techniques is usually simpler and faster, but will it result in overall project success? If time or budget is tightly restricted, it's probably best to stick to established methods.

The next section will address the scenario of completely replacing an existing building or piece of software, where the usage needs to continue as normal.

The Show Must Go On—Complete Replacement

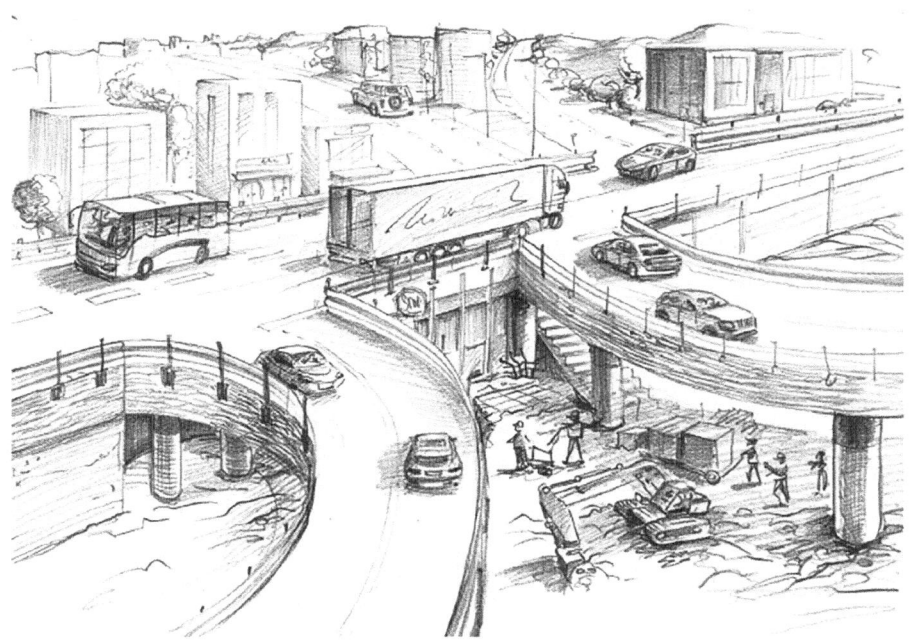

Figure 5-3. *The show must go on—even when the whole structure is being replaced*

In the many cases where a new structure will completely replace the existing one, the current usage of the building still needs to continue during the construction work (Figure 5-3). The replacement process therefore needs to be done gradually. The recent major overhaul of the main crossing in our town center is a case in point. Initially, half of the old structure was kept in use, while the other half was being rebuilt. All the traffic was temporarily redirected across half the space. Then the same was done for the second half of the road crossing. Enabling the functionality of a structure during building work of this kind naturally adds an extra level of complexity and planning.

Alternatively, an entire building may be demolished and another constructed in its place without any kind of overlap. Any items of value inside the building must first be located and retrieved, and this can be a tedious task. In software, the existing system may contain important data that needs to be moved to the new system. This data may include customer documents such as contracts and legal waivers as well as user settings and so on. The old software application that needs replacing is referred to as a *legacy application* or as *legacy software*. Depending on the age of the software and the support level upheld, the know-how among employees may be scarce. If this is the case, you'll find that extracting this data could be an awkward and time-consuming task.

The existing system can also be a valuable source of information for the team building the new system. Understanding how the current users work is a great help in identifying useful features and mandatory data. Additionally, the users of the legacy system can give valuable input by describing any weaknesses they've encountered.

Ideally, the project team should sit with users to observe how they engage with the software.

This approach gives teams a much better insight into user requirements than any meeting-room assessments.

Finally, make sure that *all* the required features of the old system are replaced—otherwise, you might end up with both the legacy system and the new one in use at the same time. This outcome will double the cost of maintenance and possibly of the license fees as well.

Next, continuing the theme of architecture, we'll compare the processes of renovating and extending buildings to the work of updating software and adding new features.

CHAPTER 6

Renovating and Extending

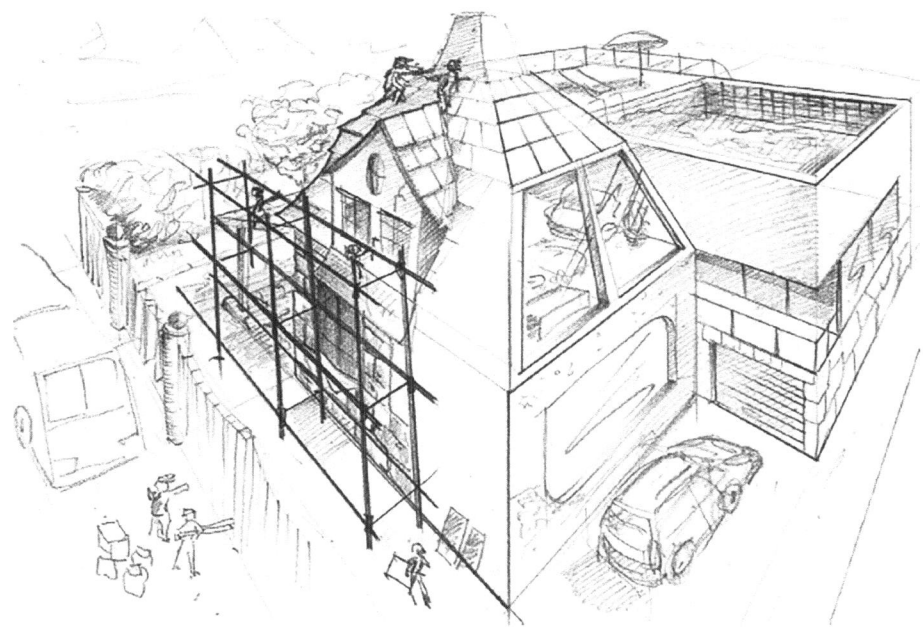

Figure 6-1. *Renovating and building an extension*

A beautiful older house may need some structural repairs and new appliances (Figure 6-1). A house left to decay for too long, on the other hand, may need demolishing. The extent of renovation work needed can

© Jonathan Peter Crosby 2023
J. P. Crosby, *The Business Manager's Guide to Software Projects*,
https://doi.org/10.1007/978-1-4842-9231-0_6

vary enormously—from replacing the kitchen to hollowing out the entire building and retaining only the shell.

For centuries it was common practice for people to build an extra floor on top of an existing building when they required more space. This practice is sometimes still followed today in cases where the underlying structure can support the extra weight. For such projects, shared utilities such as water, heating, and electricity need considering. Is there enough water pressure to reach the new level? Is the heating system powerful enough to heat this additional space? In software, a team adding completely new features also needs to look at the existing system and reutilize the shared services. The existing functionality cannot be ignored. The duplication of functionality could result in diverging logic and lead to higher costs for future changes. Nevertheless, I've seen teams making such duplication a few times, particularly when a new development team started adding new features. This situation can occur if no development work is done to a software system over a long period. The original developers may have been reassigned to other tasks or may have left the company.

When changing an existing software system, developers need to consider the current architecture. The software development team should possess knowledge of the existing technology in use and understand how the current system was built.

When renovating an old house, having access to any plans of the house could save you a lot of time. In software, retrieving and analyzing existing documentation can be a great help too. Some software developers prefer to start a project afresh and fail to take advantage of this valuable information—even though doing so may increase the time and cost.

During extensive building renovations, occupants normally vacate the building, but an existing software system still needs to function throughout the project. The team usually makes a copy of the software and works on changing and extending the copy only. A common term for this copy is the *development environment.* The current version in use by real users is usually called the *live* or *production environment.* When sufficient attention is given to understanding the existing system, this approach should allow for a seamless changeover to the updated version.

In the next section, we'll see how in software, as in a building renovation or extension, you should always expect the unexpected.

Unforeseen Work

Depending on the age of a property, a building may have seen various renovations and extensions since its initial construction. Unforeseen problems such as rotten timber beams, termite-infested wood, or faulty wiring may be discovered during the renovation. Such unanticipated work is naturally a major cause of extra costs and unexpected delays.

A lot of unforeseen work can also appear in software "renovation" projects. The development team will need to retain many of the business decisions built into the software over the years. Ideally, the *business logic* and business data will be stored centrally and clearly named. Software developers may not know the quality of the code until deep analysis is done or the actual programming begins. Individual cleanups alone may not take long. Software with a lot of chaotic code can account for a large and sometimes unexpected part of the overall effort though.

Unforeseen work in construction can throw a project right off the timeline and budget, especially since detailed information on the original work is unavailable. As a precaution, an additional 15–20 percent is often added to a renovation project budget to cover the cost of such unforeseen work.

You'll find that a similar approach is good practice in software "renovation" projects too. The amount of unexpected work will depend on the documentation of known issues and on how well the current team knows the code in use. If the development team cannot easily identify the quality of the existing code, a preliminary analysis project will allow for a more precise estimate.

High-quality code and up-to-date documentation will naturally facilitate the process of adding new features as well as the smooth day-to-day running of the software. Also, as part of a software "renovation" project, some clearing up is necessary. Neglecting this task of cleaning up will result in an increasingly burdensome technical debt, which is the topic of the next chapter.

CHAPTER 7

Technical Debt

It's not just the parts you can see in a piece of software that matter. Cleaning up critical technical debt can be at least as important as adding new features.

> *Technical Debt is a wonderful metaphor developed by Ward Cunningham to help us think about this problem. In this metaphor, doing things the quick and dirty way sets us up with a technical debt, which is similar to a financial debt. Like a financial debt, the technical debt incurs interest payments, which come in the form of the extra effort that we have to do in future development because of the quick and dirty design choice. We can choose to continue paying the interest, or we can pay down the principal by refactoring [improving code structure] the quick and dirty design into the better design. Although it costs to pay down the principal, we gain by reduced interest payments in the future.*

> —Martin Fowler [Fowler Martin 2003]

In one software solution I was responsible for, we reserved 5 percent of the budget for each release to clean up technical debt—this kept the system healthy and more adaptable to change.

© Jonathan Peter Crosby 2023
J. P. Crosby, *The Business Manager's Guide to Software Projects*,
https://doi.org/10.1007/978-1-4842-9231-0_7

Clearing Up the Mess

Figure 7-1. *Looks fine from the outside, but it's actually a big mess inside*

In extreme cases, technical debt can be compared to the house of someone who is very messy (Figure 7-1). The place may look fine from the outside and may even function largely as expected, yet one look inside will reveal the chaos. That's what happens when the occupant keeps piling things up and never gets around to clearing up. If they want to move a piece of furniture, for example, clearing a path first could be a major task in itself.

Once the development team identifies any technical debt, it needs to be addressed.

Every project acquires technical debt over time. Therefore, you'll find it well worth the time and effort to pay back some of this debt in every project.

Three Heating Systems— Consolidating Redundancies

Consider a building that was previously owned by three different businesses concurrently. It had three separate heating systems. Now this building has a new owner who will use it for one company. To save costs, they want to operate one heating system. Is one of the existing systems powerful enough to heat the whole building, or will all three need replacing?

A similar scenario can present itself in software systems following a company merger or where the once separate units within a company are now consolidated. More than one system could be holding customer data. Combining these systems into one would remove a lot of complexity and duplication. Initially the merger would be a challenge both for the IT and business teams, as the past decisions of all systems would need to be consolidated. Postponing this task, however, could lead to expensive workarounds, poor data quality, and much more work in the future.

In her bestselling book *Why Simple Wins*, Lisa Bodell gives some sound advice on tackling complexity:

> *To improve our work, we have to move from a mindset of "more" to a mindset of meaning. Eliminating elements of our work can enable us to focus and achieve our goals more quickly and effectively.*
>
> —Lisa Bodell [Bodell 2017]

The Leaning Tower of Pisa—Adding New Features

Imagine adding new levels to the Leaning Tower of Pisa before ensuring that the structural work and foundations were strong enough to support them (Figure 7-2). Not that anyone would actually want to extend this beautiful monument—but working on exciting new projects that everyone can see attracts much more attention than any underground work. Securing the foundation will ensure that many future visitors will also enjoy the tower.

Figure 7-2. *Don't just focus on new features*

Adding new features in a software project without strengthening the inadequate foundation is a common mistake.

Understandably, product managers and sponsors are usually keen to provide users with exciting new features. They may want every dollar to be spent on parts that customers, sales, and management can see and use. That means it's often hard to keep any maintenance work as a high priority on the to-do list. Also, if the project budget is decreased for any reason, the technical improvements are frequently pushed down to a lower priority.

If you were planning to build an extension to your house and knew that some of the water pipes were rusty, you'd replace them first. Otherwise, the rust would quickly spread to the new pipes in the extension. The importance of addressing such known issues at the outset, both in construction and software development, can't be overemphasized. To avoid major problems, always pay sufficient attention to the functionality and health of a system. If you risk having faulty software, you may even have to replace the entire system at some stage. Having to rewrite software from the ground up is a lot more costly than maintaining a healthy code base.

Too often have I seen companies and projects spend years focusing entirely on new features while neglecting the other important tasks. Failing to provide a sufficient budget to properly maintain the background work means you'll accumulate many problems. A problem will eventually become so large that the work cannot be financed as part of a new feature project. It may require a separate maintenance project—preventing any features from being added until the cleanup is completed. Having a clean system landscape will enable your company to adapt to change more easily. It will also save you the headache of troublesome legacy systems.

Another aspect of background work involves documenting changes and keeping these records for future reference.

Preserving Know-How

In the canteen of a client, decorators painted over a ceiling without knowing it was an acoustic ceiling—tiny holes in the surface helped absorb the noise. The following day, complaints poured in about the high level of noise. The canteen had been built over 15 years previously, and the knowledge about this special ceiling had been lost. The fix necessary for reducing the noise levels ended up taking more time and planning than the painting work itself.

Keeping well-documented records of such infrastructure details clearly makes a lot of sense for everyone involved in a situation such as this. The same is true for software projects. The effort required to document and preserve such information—as well as ensuring it is easily accessible and kept up to date—is very cost effective. It will also mean fewer interruptions to the workflow.

In the following chapter, I'd like to address the topic of maintenance. By placing a strong emphasis on this area from the start and throughout the software project, you'll save yourself no end of trouble and expense.

CHAPTER 8

Maintenance

Once the construction phase of a commercial building is complete, the client gets the keys, and the building maintenance team, which will have been involved throughout the project, takes charge. In software projects, a little foresight on the part of the project team will minimize the cost and effort of future maintenance—this will benefit both the business and IT departments. It will allow the smooth running of the software on a day-to-day basis following the launch.

The software architecture will be a major factor in the amount of maintenance required later. Additionally, the code itself can also require maintenance work. Code bugs are much easier to fix than any conceptual design issues, though. Unfortunately, the topic of maintenance is often little more than an afterthought. Some evidence indicates that for every dollar spent on feature development, another ten are spent on maintenance over the whole lifespan of a system. Companies aware of this tendency naturally place much more emphasis on maintenance. The quality of support and speed of issue resolution quickly offset the maintenance costs, saving the end user's time and promoting their acceptance of the system as a result.

© Jonathan Peter Crosby 2023
J. P. Crosby, *The Business Manager's Guide to Software Projects*,
https://doi.org/10.1007/978-1-4842-9231-0_8

Swimming Pool Water—Incomplete Data Loads

Figure 8-1. *What can be done if the pool is only half full?*

A swimming pool needs regular maintenance to remain usable, but imagine a pool that's completely drained and refilled on a daily basis (Figure 8-1). If on some days the pool is only filled halfway due to an error, it will be unusable and even dangerous if anyone tries to dive in. The maintenance team will need to investigate why this is happening. We can compare this malfunction to a scheduled data load in a software application that receives data daily. Decisions will need to be made in advance on how to handle the situation if the data load is incomplete.

Should yesterday's data be used until all today's data is available? Should there be a partial data load? Who is affected by these decisions and in which way?

Finding a plastic bag in the swimming pool would be another problem—it could block the pool drainage system. The bag, which would obviously need to be removed, can represent some bad data that is delivered to a system. Can just a small amount of bad data cause the system to stop the whole data load? In some cases, it can. Is the delivery of external data being validated? Your team should define measures on how to handle data delivered in an unexpected format.

As incomplete or bad data issues are very common, all stakeholders should be made aware of this topic. You need to define sensible data guards and find ways for your business teams to continue to operate in such cases. The development team can then ensure that the system handles data well and streamlines the user's daily workflow.

Occasional Maintenance Tasks

Not all maintenance tasks are ongoing—some jobs need doing only occasionally. Think of tasks such as clearing leaves from an outside drain or changing the batteries in a smoke detector. Neglecting these simple tasks could lead to serious consequences later. The key here is to ensure that such tasks are monitored and scheduled over long timespans.

Many software issues arise from such relatively rare cases too. You may need to add a new bank holiday to a human resources time-tracking system, for instance, or introduce a new currency to a financial system. Although such changes are rare, the data can prove troublesome because software is often designed to cater only for the most common cases. Establishing a process for maintaining data that rarely changes is an area that is often neglected. In many systems, such data is added at the beginning of the project. Thinking ahead at this stage means allowing changes to be made for cases that will crop up in the future. The team needs to define a concept and create screens for adding variations later to the rarely changing data. If, for cost reasons, you neglect this step, there'll probably be repercussions requiring support over the years that follow.

I hope the comparisons with physical construction have given you some insights into software development. In the next chapter, I'd like to clarify the differences between the two fields and show how some of the construction metaphors are only rough equivalents designed to aid understanding.

CHAPTER 9

Differences Between Physical Construction and Software Development

So far, we've focused on the similarities between these two types of work—but it's also important to understand the limitations of these analogies. In software development, the construction metaphor is sometimes criticized for overemphasizing upfront planning. For the most part, building construction will not begin until the architectural plans are complete and building permission has been granted.

Most developers now favor the Agile approach for the adaptability and flexibility it gives. Agile also allows teams to better handle late changes during a project. Traditionally, however, software developers have used the rigid structure and planning of Waterfall methodology, which is comparable to the approach used in construction. Waterfall is, nonetheless, still used in some projects today and will remain necessary for certain types of projects.

Another difference is seen in the job roles. The role of architect, which exists in both professions, is one example. In physical construction, the

J. P. Crosby, *The Business Manager's Guide to Software Projects*,
https://doi.org/10.1007/978-1-4842-9231-0_9

architect controls the entire project and may even decide on the small
details. In software projects though, the development team can make
many decisions by themselves. Furthermore, some teams are now leaving
behind the classic architecture approach and moving toward a more
flexible collaborative development among themselves instead.

Other jobs such as bricklaying are not at all comparable to software
development either. If the construction work has been delayed, additional
labor will speed things up—a bricklayer could begin working almost
immediately with very little instruction. The situation is quite different
with the developers building software, though. You can't simply bring in
additional developers to help ease the workload. A new team member
first needs to learn about the project specifics and some of the business
domain knowledge, initially slowing down the rest of the team.

Throughout the construction process, the home buyers can inspect the
building site from the ground up. They can see where most of their money
is being spent and can check for mistakes themselves. In software, unless
they can understand code, the business team is limited to reviewing the
visual part.

Another difference in the two types of work concerns duplication.
The code and components used in software development can easily be
copied and reused. Software designed in a modular fashion lends itself
particularly well to reuse. The login module, for example, could be used
again in the next project. The construction plans for houses can, too, but
brick-and-mortar buildings are somewhat different—the actual physical
parts cannot be utilized elsewhere without tearing down the original
building.

As far as security is concerned, the measures you'd take for a small
house would be limited to that specific building. A security lapse even
in a small website, however, could have repercussions on any connected
devices worldwide.

A further core difference between construction and software
development is in the lifespan. Some buildings date back thousands of

years. The challenge to create and run compatible software for the future
is another matter though. For one thing, the rapid advances and frequent
introduction of new standards can quickly make devices and software
incompatible.

Every metaphor has its limitations, and it's therefore important to
distinguish between the commonalities and differences. In my experience,
construction metaphors do help non-techies to picture some of the core
concepts behind software development. The advantages far outweigh
the downside of anyone misunderstanding parts of a metaphor. When
in doubt, please consult your development team to gain a deeper
understanding of a technical topic.

CHAPTER 10

More Metaphors

The collection of metaphors in this chapter moves out beyond the construction theme. These will help give you a wider picture of the various facets of software projects and applications.

> *A metaphor is a device for seeing something in terms of something else. It brings out the thisness of a that or the thatness of a this.*
>
> —Kenneth Burke [Burke 1945 p. 503]

Here's an example to get you started: Some software applications can be compared to supertanker ships. These ships respond slowly to steering and take time to change course. Any change of course also requires careful planning. Supertanker software applications, often referred to as legacy systems, are just as inflexible, hard to change, and expensive to run. Although your software project may be relatively small-scale, the team should check early on for any dependencies the software might have on supertanker applications. The inability of these applications to change course quickly could cause delays in your project.

Now let's look at the high value of kitting out the team with the appropriate tools and equipment.

© Jonathan Peter Crosby 2023
J. P. Crosby, *The Business Manager's Guide to Software Projects*,
https://doi.org/10.1007/978-1-4842-9231-0_10

Dentistry—Choosing the Right Tools

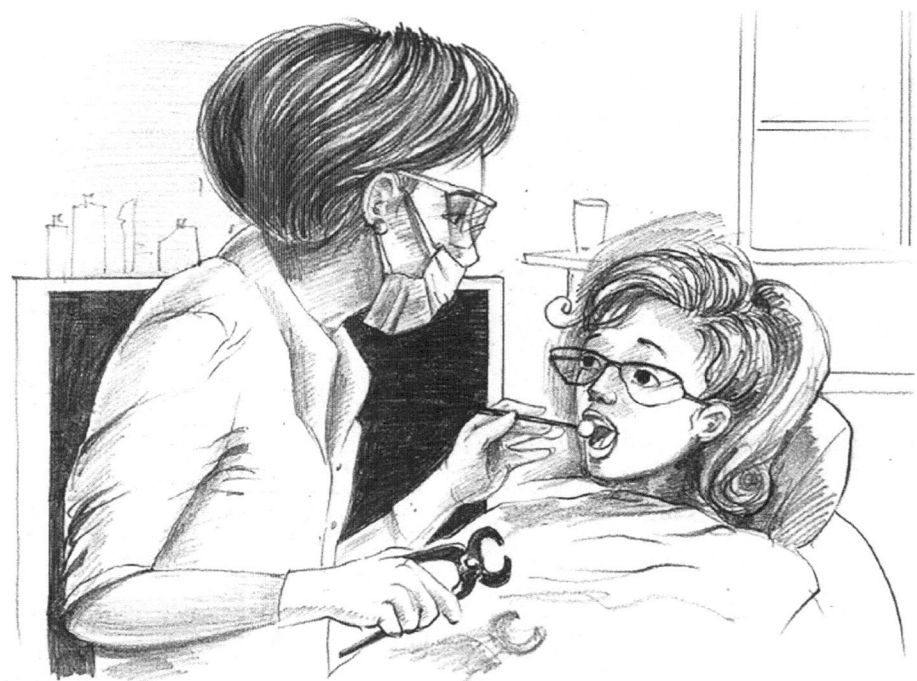

Figure 10-1. *The more precise the tool is, the longer the work will take*

If your dentist produced a large pair of pliers to work on your teeth, you'd be off like a shot (Figure 10-1)! We naturally expect the dentist to use the tools specific to the task. Selecting the right tools is crucial for any work we do, and software development is no exception.

Software tools come in all shapes and sizes. Some come out of the box with many useful features and are simple to configure. While suitable for many standard tasks, these may not allow for any finer adjustments, however. To identify the best type of software for a task, the developers should consider: Has this task been done before? Is there already a toolkit that can be configured according to our needs? When there are dozens of toolkits that fulfill a similar objective, the next questions will be: Which one meets our needs best? Shall we use open source or licensed software for this? Is there a wide community using and supporting this toolkit? In cases where developers find no suitable software available, they will then consider developing software from scratch. Although creating new software usually takes a little longer, it does minimize dependencies and allows for complete customization.

To illustrate the importance of making the right choices for your project, imagine that you bought a low-resolution digital camera and took a few pictures. It would be almost impossible to get a higher resolution or to add more detail to the picture on your computer. If you used a high-resolution camera, however, you could easily compress the image later, depending on the purpose. The choice really depends upon your needs, but having the option to work on a granular level can give you a lot more possibilities. Having such flexibility typically does cost more, though.

The granularity of standard software may be insufficient to reflect the most detailed elements of the business. It can be very hard to add another layer to meet your project needs. It's therefore essential to make sure the tools you choose are the right ones for the task.

Next, I'd like to look at how more work is required when replacing older and highly dependent IT systems.

Deep Roots—Linked Systems

Conditions permitting, the roots of a tree will gradually grow further and further into the ground. The older the tree gets, the stronger and more deeply the roots will go (Figure 10-2). Digging out the tree will naturally become more difficult as time goes on.

Figure 10-2. *An interwoven system is like the deep entangled roots of a tree*

Similarly, with IT systems that expand over time—especially by connecting with many other systems—you'll find any replacement or removal of such legacy systems very work intensive. Every new dependency to another system is like the roots of the tree growing a little more deeply into the ground. As many of the features and business decisions built into the software over the years will still be relevant today, replacing such an interwoven system is no simple matter. Determining which data needs to be retained will usually mean coordinating various stakeholders and teams. The typical layout of today's IT systems will resemble a spider's web rather than the roots of a tree.

Having data flowing across many systems can be another source of trouble if the delivery is unreliable, for example. We'll investigate this data flow in the next section.

Gently Down the Stream—Data Flow

Figure 10-3. *Data flows through a company's systems*

Just as water can flow into a river from any number of springs and streams (Figure 10-3), a website or another system can receive data from many different systems in the background—and all without the user's knowledge. Just like the water in a river, data typically flows across systems in one direction, but there are also cases where data flows in two ways.

Downstream applications are strongly affected by any issues in the upstream applications passing data. Delays in making changes or fixing data quality issues can be very painful, causing substantial difficulties in the development and testing phases of your project as well as affecting live systems. This drastic outcome may be due to the other team responsible for sending the data focusing on their own features in the application for their users. The team may give a lower priority to sending data to the recipient system. Any delays in an upstream application will cascade down to all the recipient systems. You'll find that this effect is particularly noticeable where a chain of systems is waiting to test new data.

Measuring Quality

Consider a narrow river that runs through a town, where various factories on the banks draw and pump back water along the way. If the quality of the river water is unaffected when it passes beyond the town, then the factory is not polluting the water. Likewise, a simple way to monitor the health and reliability of a company's systems is by monitoring the downstream applications and reports. Existing downstream applications that consolidate data from many systems will offer valuable information on which sources are reliable and stable. The data can also reveal which teams work efficiently and fix issues swiftly. Having worked on various downstream applications, I know that this rich source of very useful information is largely untapped. At the same time, at most of the companies I've worked for, there was a considerable discrepancy between how teams *actually* worked and the impression the management had of how things were done. Some systems were incredibly well-run with good communication and strong technical skills, whereas other systems were unstable and the team often forgot to communicate important changes. Using this information from downstream applications would help address problematic systems and improve processes, even across departments.

Speaking about this topic a while ago, a previous colleague half-jokingly made the memorable remark, "We find the worst system in our company and then make it strategic."

Now we've looked at the main concepts behind software projects, it's time to roll up our sleeves and move on to the more practical side of the work.

PART II

Practical Guide

CHAPTER 11

The Practical Side

Having learned many of the conceptual ideas behind software projects, you'll be ready to see some action. In the chapters that follow, I'll help you turn this knowledge into concrete contributions. We'll look at the whys and hows behind some common project scenarios.

When implemented correctly, technology should help the business focus on their core competencies—gathering the relevant information and displaying it fast and in the right form. Further, technology can help reduce complexity and simplify processes for the user.

Please note that project tasks don't necessarily follow a strict chronological order. Some tasks may run in parallel or might be unnecessary in smaller projects. While the main focus here will be on the areas where business users have the strongest impact and influence, I'll also give you a brief insight into the other typical project phases.

In Part 2, "Practical Guide," there is a strong emphasis on identifying the requirements. Many of the business tasks lie in this preparation stage, when stakeholders make the crucial decisions. These decisions will influence the subsequent steps of the project and are therefore vital for overall project success. Now that the value of close collaboration between the business and IT teams is widely understood, it is much more common practice in software projects. The testing stage at the end of the project again requires the close involvement of the business departments. The testing helps ensure the acceptance criteria have been met.

© Jonathan Peter Crosby 2023
J. P. Crosby, *The Business Manager's Guide to Software Projects*,
https://doi.org/10.1007/978-1-4842-9231-0_11

CHAPTER 12

Plan and Prepare

Getting Off to a Good Start

The way you plan a project will have a huge impact on the outcome. You'll need to carefully consider many important factors. By establishing good communication between the various professionals involved, you'll get your project off to a very good start.

There's also a growing need for upper management to involve themselves more with information technology and software projects. Accordingly, I've written specific sections with executives in mind. Wise decisions made at the board and upper management levels will benefit both the business and IT sides enormously.

This chapter will give you deeper insights into how software projects work. Although some points may just sound like common sense, you'd be surprised at how many projects struggle or fail because these basics are neglected.

You'll find that establishing clear goals and a rough road map at the outset is very helpful. The first and most important task for the business stakeholders is to define clear business requirements. Once these are defined, the project team will later break them down into smaller, more manageable parts.

© Jonathan Peter Crosby 2023
J. P. Crosby, *The Business Manager's Guide to Software Projects*,
https://doi.org/10.1007/978-1-4842-9231-0_12

What exactly do you want to achieve with this project? What is the underlying business objective independent of the current systems? How will you yield the most business value?

Most software teams and companies adopt the *Lean and Agile* approach these days. Many books are available that explain how to organize these teams for maximum efficiency. While taking this method into account, I try to focus on software projects in a broader sense, regardless of the project methodology used.

What steps does a software project involve?

Whichever software project methodology you choose, it's usually best to consider the steps listed in the following. In a project using the *Waterfall* methodology, each step is executed once. With an *Agile* approach, the project is broken down into smaller iterations that, in turn, also contain most of these steps.

Steve McConnell, author of *Code Complete,* describes how "Researchers have now identified numerous activities that go into the complicated process of software development. These include:

- Problem definition

  What goal shall be achieved?

- Requirements development

  What services shall the software deliver?

- Software architecture, or high level design

  What is the overall structure of the system?

- Construction planning

  Which order shall the services be built in?

- Detailed design

What specific internals shall be built?

- Coding and debugging

  The actual creation of the software

- Unit testing

  Testing the smallest parts of code

- Integration testing

  Do the various parts of software work together?

- Integration

  The actual step of merging the software parts

- System testing

  Does the software work with other systems?

- Corrective maintenance

  Restoring a system fault to fully operational"

—[McConnell 2004]

These steps will be performed by a whole team of specialists, especially in larger projects. "Appendix A: Collaboration" gives you an overview of who does what. Business professionals will largely need to focus on the key areas of problem definition, requirements development, planning, and testing.

Executives have a key role to play in enabling teams to perform at their best. In the following pages, you'll find some examples and suggestions on how executives can support a project.

Executive Support

Every one of my clients over the last 20 years placed a high importance on IT and on their business systems. For some reason though, they often tended to regard the IT department as a back-office division.

> *Sadly, an IT department that has been run as a cost center for years or decades while valuing stability over change, is not well prepared to provide business agility and flexibility to the business.*
>
> —Gregor Hohpe [Hohpe 2016]

Studies show, however, that companies benefit most from technology when the IT teams are treated as equal partners.

Meanwhile, established tech companies and software-driven start-ups have been entering markets outside the IT sector, where they are competing and taking market share. They do so swiftly, aggressively, and with a high focus on customer needs. Most often, these new competitors treat their IT specialists as first-class citizens rather than engine-room workers and can therefore attract top talent. Any leaders ignoring this trend, still clinging to the methods that were successful yesterday, will probably face tough times. Their companies may be unable to adapt to change fast enough. *Are you a digital predator or digital prey?*

"We all know the stories: Netflix versus Blockbuster, Amazon.com versus Borders, or iTunes versus the entire recording industry. These cases are just the beginning of the larger shift, where companies that are able to innovate quickly with software will out-compete traditional market leaders" [Forrester Consulting 2013].

In many companies though, the IT specialists and even IT managers don't have enough influence to establish better working structures. Instead of simply granting project budgets to the business departments, it's time for executives to change this secondary status of IT and transform the way that teams work together.

Digital transformation is not an IT project; it's about changing the whole organizational culture.

Members of upper management need to understand the importance of technology for their company and for the types of decisions they need to make. Many companies have the required knowledge in-house already, yet it may have been left untapped so far. Having a few individuals around who understand the concepts and best practices in software projects is useful. The biggest gains, however, come when management and the project teams understand the software fundamentals. In tech companies this situation is naturally the case. Successful technology projects are a team effort that goes far beyond the IT department. The management should therefore lead the way and be a part of the transition themselves.

Executives need to ensure that mission-critical projects are not delayed because of less important ones. Putting other projects on the backburner for the time being can help teams focus on what matters most.

Another critical topic is poor data quality—this issue can affect systems across the entire organization. Therefore, the decision to tackle this should be made at the executive level.

Technology, together with the teams enabling it, takes up a considerable proportion of the overall operation costs. Yet technology done right will save costs, improve quality, and increase your return on investment. It will also leverage scale and give you an edge over the competition. That's why businesses worldwide are investing in this area. No one is considering reverting back to manual methods.

Traditional companies are now starting to adopt the best practices that have been used for years by the leading tech companies. Amazon and Netflix release software updates, new features, and bug fixes thousands of times every day. Although your company may not need such frequent releases of software, you surely need the capability to react quickly to stay on top. Trying out different strategies on a subset of customers to keep your edge is becoming an essential business strategy.

Work Culture

For thousands of years, China was one of the world's greatest innovators. They invented papermaking, the compass, gunpowder, and printing among many other things. Railway expert Jun Cui feels that these days though, the Chinese are far less innovative than they could be, however, and blames the hierarchical Chinese culture.

As Jun Cui says

> *Since Confucius times, the children are taught don't talk back to parents, don't talk back to your managers, your superiors. Even if you have a better idea, you better keep it to yourself, because you will make your superior look stupid, because the superior should be the smarter one. But that isn't true.*

—Jui Cui [Nottaris et al. 2016]

Although many of Confucius's teachings, such as "What you do not wish for yourself, do not do to others," are still relevant today, I fully agree with Jun Cui on this point. Literally anyone can come up with a great idea. More recently, a lot of innovation has come from Chinese IT and technology companies, which I believe could be due to these younger industries' use of a more inclusive approach to innovation. For instance, the tech companies Alibaba and Tencent control over 80 percent of China's trillion-dollar online and mobile payments market—leaving the traditional banks with a minority share.

History has shown us that some great inventions have been discovered by people with no formal education in that field. One of my favorite examples of such a person is the self-educated English clockmaker John Harrison. He solved the problem of calculating longitude at sea by making an accurate ship's chronometer, thereby advancing navigation considerably and making voyages much safer. The Board of Longitude, the official body that had offered a monetary reward to the person who solved this problem, was reluctant to present the full amount to a layperson, however. They had expected the solution to come from a respected astronomer or prominent scientist. Harrison subsequently appealed to King George III, requesting him to intervene. Parliament eventually paid Harrison the reward in full shortly before his death, but he was never officially recognized for his achievement.

Another example is the electrical engineer Harry Beck. Experimenting in his spare time, Beck came up with the definitive design for the London Underground map in 1931 (Figures 12-1 and 12-2). His simplified map looked more like a diagram with streamlined shapes and lines. The design was originally rejected, but he persisted in getting his work recognized. Eventually a few hundred copies were printed. These proved so popular that a reprint of 700,000 copies soon followed. Since then, many other transport systems around the world have adopted the same style of map.

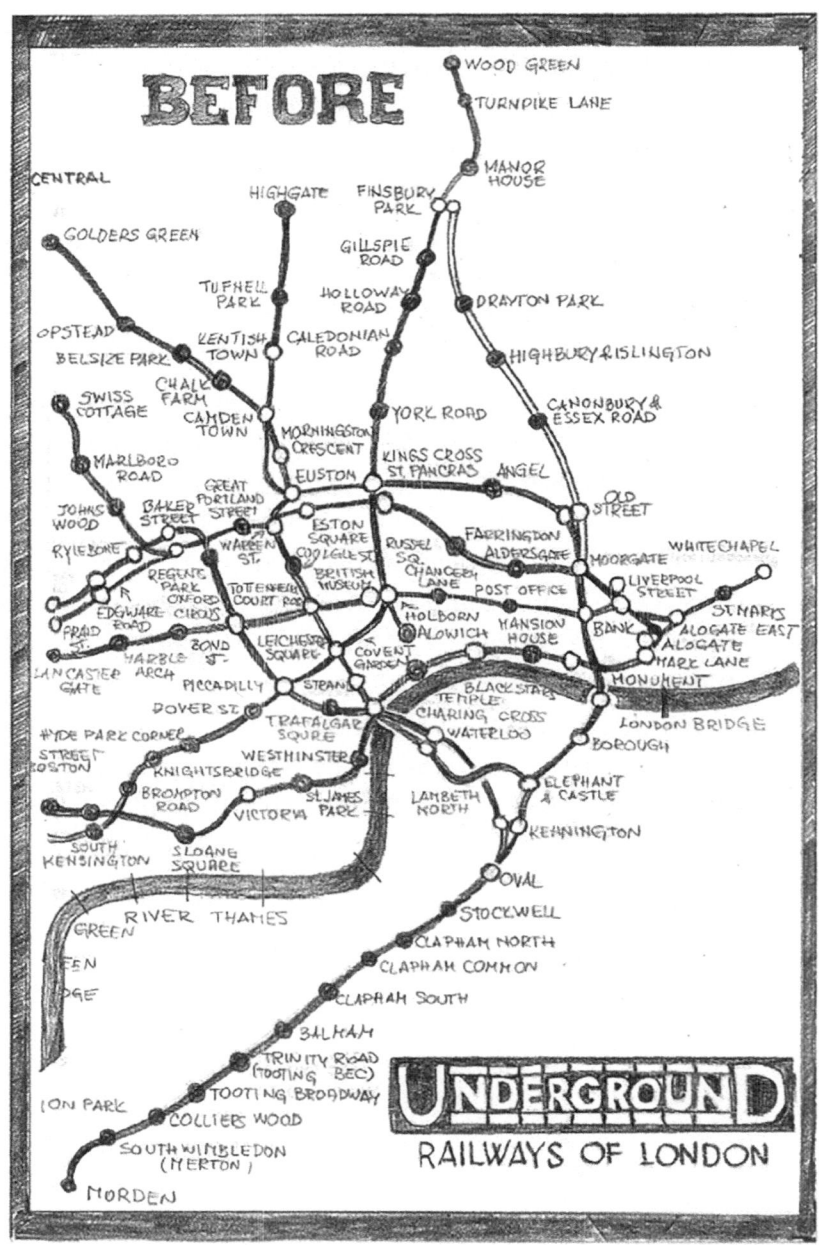

Figure 12-1. *The original map*

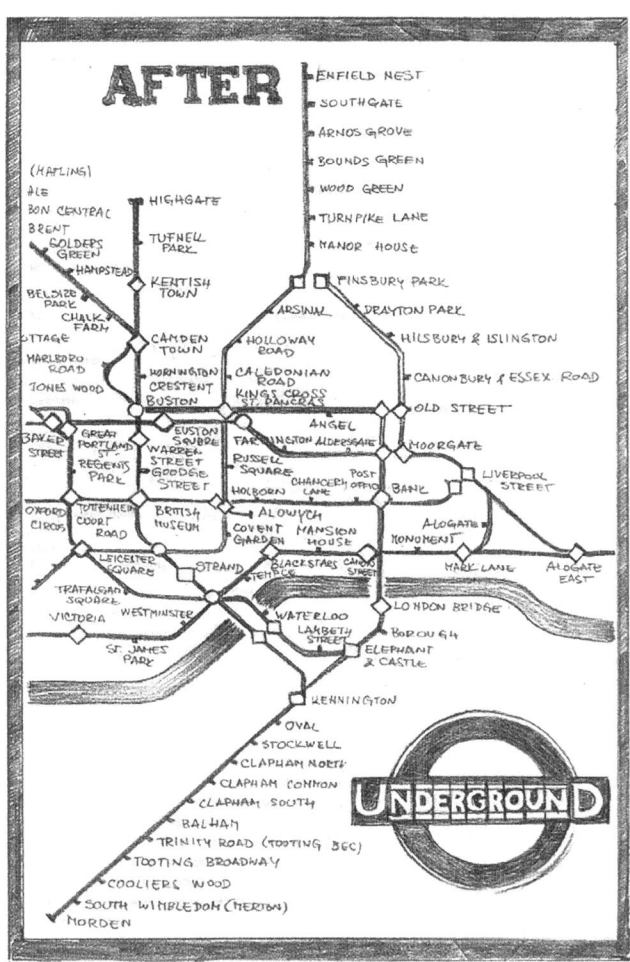

Figure 12-2. *Harry Beck's innovative map design*

The examples of these two inventors demonstrate that anybody can come up with an amazing innovation, not just a few selected experts.

Ideas need to come from the lunch room and not from the boardroom.

—Benno Dorer [Dorer 2017]

By encouraging the free flow of ideas, you'll add value to your organization. Many companies, as well as individuals, still avoid making bold decisions and choose the path with the lowest risk instead. Such an outdated approach means that some large corporations need to purchase innovative solutions from elsewhere. Disciplined and prudent risk-taking is at the heart of sustainable success. Therefore, empower your employees to make decisions. Give them the opportunity later to reflect collectively on both the good and bad decisions they've made. By doing so, you'll help create a powerful corporate culture. In fact, research shows that employees who feel empowered at work perform better, are more committed, and show higher job satisfaction.

In nurturing the optimal work culture, the aim is also to encourage an atmosphere of openness that allows for failure.

> *Fostering an atmosphere that doesn't allow for error, simply makes people defensive. They don't try things that may turn out badly.*
>
> —[DeMarco & Lister 2013]

I also strongly recommend providing team members with continuous feedback to highlight their strengths and weaknesses. Managers and project leaders giving sports coach–style feedback find it far more effective than any annual reviews.

Working Better Together

The Phoenix Project [Kim, Behr & Spafford 2013] is a book full of amazing insights on improving the ways people work together. Taking the form of a novel, it follows the processes within and around the IT department of a company as they transform these processes from bad to good. The story line is great, and the book is easy to read, giving technical details in only a few places. I strongly recommend adding this book to the must-read list for managers. It's especially useful in illustrating how the different company departments can work together more effectively.

One of the crucial points to come out of the book relates to *cooperation*. To promote cooperation among employees, management must dispel the *silo mentality*. This term refers to how departments or teams work in their own interest, without focusing on how teams within a company work well together.

At the same time, IT teams and managers alike often complain about the increasingly heavy load of internal regulations passed down to them. Fulfilling every one of these would bring the real work to a complete halt. Of course, companies must act in accordance with the law, but in my experience, reams of internal guidelines, regulations, agreements, policies, and so on are produced, which far exceed the legal requirements. Some companies try to avoid the repetition of tiny mistakes by overloading the processes with rules, rather than addressing the issue with training or through simple communication.

To simplify things, encourage the internal risk, compliance, and security teams to assess the risks per department or even per team instead of simply presenting the entire list of regulations and the like to everyone. Also, prioritize the items to ensure that the biggest risks are addressed first.

Could your company separate these risks and regulations into rules (essentials) and guidelines (targets)?

As Erik (the board candidate) says to John (a senior executive) in *The Phoenix Project*, "You win when you protect the organization without putting meaningless work into the IT system. And you win even more when you can take meaningless work out of the system" [Kim, Behr & Spafford 2013].

The framework for project management needs to cater for a large variety of projects. Nearly all companies will deal with highly complex projects as well as simple ones, and the project framework needs to cater for these varying needs. A company therefore should allow for some flexibility in the structure and financing of the project. For a very small project, for instance, the team should not be obliged to fill out a pile of documents that are not actually applicable.

The way a team operates will also be strongly influenced by the software methodology they use for a particular project. Most software projects today use Agile methodology. Tasks are divided into short time-boxed windows of work. The business and IT teams meet at regular intervals, frequently reassessing and adapting the planning. As stated in the Agile Manifesto, some of the key concerns involve interactions, collaboration, and democratic team decisions [Manifesto for Agile Software Development 2001].

As the product is at the core of every project, companies benefit more from pursuing product strategies or even whole business strategies rather than actual projects as such. Many software solutions involve an ongoing process rather than a finite project because technology is constantly evolving—creating value for customers. Focusing too much on projects will often lead to products being continuously enhanced with new features, but with no funding available to finance important cleanups or longer-term strategies.

Integrating the Teams

At one company I worked for, the IT department responsible for the mission-critical software applications was fully integrated with the business departments. We didn't just merge on a hierarchical chart, but also moved together physically. This practice greatly improved communication and also boosted team spirit. The feedback both from the business and the IT departments was very positive. While this approach may not work for all areas of IT, and not for all companies, it's always worth remembering that the business systems reflect the core of many business processes. Having teams integrate can make a lot of sense, as both sides can benefit from sharing their know-how.

At the other end of the scale, I recently heard of the adverse consequences that resulted from poor organization within one tech company. Here the product support and software development teams belong to separate departments. As their first common manager is the CEO, any escalation of issues from these teams quickly reaches the board level. Some of the software the company sells costs hundreds of thousands of dollars in license fees each year, and the customers of this specialist software are technical experts. At the same time, all customer support cases are obliged to go through first-line support. So what was going wrong? When these tech-savvy customers called with issues, the first-line support simply requested them to reboot their workstations. This response naturally infuriated the customers. Some of them replaced the software, partly due to their dissatisfaction with the low-quality support. The board is still arguing about this setup. Clearly, the company is failing to provide the right processes. Had it allowed these customers more immediate access to the necessary specialized support, their issues would have been nipped in the bud.

What kind of experience have you had when giving feedback to companies? Have you ever noticed a missing feature or even found a bug? Did you report the issue and get a useful response?

While the preceding example doesn't shed a positive light on the first-line support, in general, I do strongly appreciate the work that support teams do. Having worked in support teams early in my career, I see the value of a company having an effective help desk. The large number of different systems in a company that support teams need to understand today is mind-boggling. The support team is actually a major source of valuable information that is often left untapped. As well as having a good overview of the health of company systems, they're in an ideal position to collect both internal and external customer feedback. Other ways of accessing this data are far more expensive and time-consuming. Product innovation is often the direct result of listening to clients' needs. Even if only a small fraction of customer feedback leads to actual changes, the analysis of this data is still worth every penny. The problem is that the data often doesn't reach the relevant teams or persons, especially if the first-line support has been outsourced.

Invest in Employees

Again and again, I've attended large department meetings where staff were drily informed that an old system was about to be removed. A brief announcement of this bare fact leaves a lot of questions unanswered. Many people could potentially lose their responsibilities within this system that may have been at the center of their working lives for years. They may fear losing their status as experts and even losing their jobs. The same people, however, will be expected to ensure a smooth transition to the new system. More than once, I've witnessed such specialists leaving within 6 months of such an announcement and being warmly welcomed at their new place. Due to the poor communication and lack of perspective, the specialists are gone. The decision to replace the old system may be valid from a technical point of view, but it's good to think a few steps ahead and consider the effect of such messages on the employees. The old system may need to be kept running for longer than planned, and the employees' knowledge will be useful far beyond the handover. This kind of foresight can avoid expensive recruitment, prevent the loss of critical know-how, and give the employees a new and exciting challenge at the same time.

Providing your employees with continuous training will prepare them for an ever-changing business model and IT landscape. In fact, research reveals there's a systematic underinvestment in human capital. Increased productivity will come by employees working smarter, not by working longer hours.

On the other hand, as the software tools we use continuously improve productivity, executives must make tough decisions and replace legacy software as necessary. Such action may be inevitable even when it affects the roles and responsibilities of long-standing colleagues. I've seen inadequate software being kept in use due to a friendship between an application team head and a C-level executive, for example. Such bad practice emphasizes once again the importance of facilitating smoother transitions into new roles. Having an adaptable and well-trained workforce ready for change should be among a company's primary concerns.

The invention of the tractor destroyed millions of jobs worldwide. Other jobs were created. That's been the cycle for hundreds of years now. You can't stop innovation. Trying to keep teams and jobs alive artificially by supporting out-of-date technology has never worked. Instead, by rotating jobs and providing continuous training, a company will motivate and add value to the workforce. This can only be achieved with employees who feel safe in eliminating their current role, knowing the company will provide a new one.

In tapping the potential of staff, I firmly believe in giving people of all ages a chance. Even non-IT companies have large IT departments, and it's our duty to train a fair share of young people. Simply cherry-picking experienced individuals will leave a gap somewhere else. In the mid- to long term, these investments in people will pay off for the whole economy. I've also enjoyed working with older, more experienced IT specialists too. It's a shame if talent and experience are wasted by overlooking members of the older generation. In many countries, companies can still discriminate on age without any penalties. I believe projects can benefit enormously from diversity in terms of age and in a mix of men and women from different cultural backgrounds.

In his bestselling book *The Hard Thing About Hard Things*, Ben Horowitz says

> *I learned about why startups should train their people when I worked at Netscape. People at McDonald's get trained for their positions, but people with far more complicated jobs don't. It makes no sense. Would you want to stand on the line of the untrained person at McDonald's? Would you want to use the software written by the engineer who was never told how the rest of the code worked? A lot of companies think their employees are so smart that they require no training. That's silly.*

> –[Horowitz 2014]

At one former workplace, I knew a manager who, in an attempt to advance their own career, hardly ever utilized their department's training or travel budgets. Further, they always flew economy class, despite business class being the norm. While that was fine, their cost-saving efforts in staff training were not in the company's best interests. By way of contrast, a mid-sized IT company, which placed programmers in my team, sent its employees on 5 weeks of training courses every year. These software developers were among the best I've known. Hence, I recommend that executives strictly monitor and challenge managers who do not fully utilize the training budgets assigned. Also, arranging for teams to travel to meet and exchange know-how with other remote teams is useful for improving team collaboration and enhancing knowledge.

Realistic Timelines

On several occasions, I've seen upper management imposing short-sighted "bookkeeping-style" deadlines. Imposing an early project deadline simply to avoid the cost of renewing the software license meant that one project was faced with an end date that was unachievable from the start. The deadline was set even before the main business goals of the new software had been defined. Not once were any such deadlines met. In this case, by rushing the new software project, many temporary solutions and poor shortcuts were built in that later took years to put right. This short-sighted approach also led to frustration for the end users, who had to use some features in the old software and others in the new application. To top it all, the license fees and maintenance costs had to be paid both for the old and new software for years. Clearly, none of this was in the company's best interests, not even the bookkeeper's.

I've even known executives who set unrealistic timelines intentionally. They assumed that setting a realistic one would encourage the teams to take it easy and waste time.

Failing to allow sufficient time for a project results in shortcuts, more bugs, and a terrible work morale. Presenting a challenge is fine, but it should be an achievable one.

In *The Mythical Man-Month*, Fred P. Brooks gives us an important lesson, "Adding manpower to a late software project makes it later" [Brooks 1995 p. 25]. This truism is even more relevant as you get closer to a deadline.

If you're in charge of strawberry-picking work, you can easily speed things up by bringing in more workers, as in the example of bricklayers given earlier. It doesn't work like that in software development though. When someone starts a new job that has a steep learning curve, it will naturally take time before they are fully productive. Getting a new programmer up to speed can take weeks or even months, depending on the complexity of the project. Added to that, training new team members will also slow down the existing team.

Managing Dependencies

Having often worked with systems receiving data, I know just how frustrating it can be to depend on the work of unreliable systems, teams, or individuals. Ensuring the various systems and teams work well together is also a high priority from a customer and company perspective. This priority is not always reflected in the teams' objectives, however. As their primary goals are often set to please their direct users first, the cross-system flow of data has a lower priority. Multiple IT systems may be involved in the process of opening an account for a new customer. The customer simply wishes to open the account as quickly and easily as possible. Any issues between IT systems in the background need to be resolved with the customer perspective in mind.

Solving issues across systems is often a slow and laborious process. The initial response of some busy teams tends to be that their system is working fine. This premature response obliges the recipient system's team to provide additional proof of the technical or data issue. If bad data is coming from further up the chain of systems, it may require a whole chain of teams to pinpoint the root cause. Quite often, these types of issues will necessitate IT management or team head involvement. Companies should implement ways of identifying such overarching issues. You'll find that using sophisticated monitoring tools for such interdependent systems helps identify the root cause quickly, thereby avoiding or reducing any managerial intervention.

One very common cause of issues is that little or no impact analysis is done when changes are made.

Let's say a new product category is added in System A. No one responsible for Systems B–F has been informed, even though they may need to adjust to this change too. This lack of communication and analysis can severely affect the way a company works. Fixing things in emergency mode leads to cutting corners and temporary solutions too. Keeping these affected systems running may also require considerable manual intervention. High-performing organizations take measures to minimize such scenarios and spend less time on this kind of unplanned work.

Dependencies on other teams exist in all areas of enterprises. Having low-quality, one-sided dependencies makes teams fully dependent on another team, which has no incentive to lift a finger. The effect can be devastating. In my experience, these one-sided dependencies are one of the biggest issues companies face, especially if some teams don't focus on the customer perspective. Aligning goals across teams and creating feedback loops can help improve this situation. A common way to tackle this issue in software projects is through the DevOps approach, where developers and operations teams work more closely together.

"Amazingly, the transformations are not primarily based on automation. Instead, the incredible improvements come from modifying policies around the system of work and the policies that control work in process, ensuring that there are effective cross-functional teams, subordinating everything to the constraint, and managing handoffs well" [Kim, Behr & Spafford 2013].

IT Infrastructure

One of the goals of executives is to make their companies more efficient. A simple way of achieving this goal is by investing wisely in the tools your employees work with every day. When I first started working in IT, both my home computer and Internet connection were much slower than the infrastructure at work. Today the reverse is true. My $800 computer at home is much faster than the devices my clients provide for their employees, and my home Internet connection is much faster too. In many companies, any request for new hardware is treated like a child asking for a pet crocodile for their birthday.

The tools your employees use for many hours every day should be reliable and fast. In reality, however, the cost of employees' wasted time far exceeds the hardware costs—but a gain of only a few percent efficiency is worth every penny spent. It's not just the developers and designers who require high-quality devices—I've also seen business users struggle to open large spreadsheets on their desktop devices. In fact, having decent tools not only increases efficiency but also prevents a lot of stress and frustration among employees, boosting job satisfaction and motivation instead. Studies show that employees lose more than 30 minutes per day because of slow technology.

Supporting the Development Teams

As well as providing fast devices, managers and executives can enable developers to work as efficiently as possible in other ways too. The upper management can direct the right attention to ensure the developers have quiet work areas that allow them to concentrate. The book *Peopleware* looks closely at the topic of office space and work environment for software developers:

If your organization is anything like those studied in our last three annual surveys, the environmental trend is toward less privacy, less dedicated space, and more noise. Of course, the obvious reason is cost. A penny saved on work space is a penny saved on the bottom line, or so the logic goes. Those who make such a judgment usually perform a cost/benefit study without studying the benefit.

—[DeMarco & Lister 2013]

Managers really need to understand the importance of providing their developers with a quiet work space. My own experience in one of my first jobs is an example of the frequent lack of awareness of this factor. Tasked with writing software tools for multiple help desks, I also spent time answering help desk calls. My boss asked if I could program in between calls. He was a great guy, but he had no experience as a developer. As a manager, he was used to switching tasks and was unaware of the programmer's need to concentrate for long periods without interruption.

When you think of an artist working on a painting or a composer writing a piece of music, do you imagine them answering the phone every 15 minutes? Or do you envisage them going into a trance-like state of deep concentration to work?

Automation

Another important job for executives is to identify the work processes that are suitable for automation. The deciding factor is really based on whether the outcome from a business process is predictable or not. If the outcome changes regularly, and future changes are hard to predict, the process will be harder to automate. Hence, processes that require a lot of manual work and/or those with a predictable outcome are suitable candidates for automation. Automating the writing of a journalist's articles would be much more difficult than automating the distribution of customer statements, for example.

Outsourcing

Outsourcing involves handing over internal tasks to external companies, and this is quite common for IT tasks. These companies can be grouped into *onshore*, *near-shore*, and *offshore* categories, according to their location. The first usually refers to an outsourcing company based in the same country. Near-shore describes an external company in a country nearby, and offshore involves a company located further away.

The topic of outsourcing is a delicate one—most managers and IT specialists have strong opinions on the subject. In my own experience, nearly all the outsourced software projects for business applications have resulted in a reduction in quality.

For one thing, after celebrating the short-term success of the project, the outsourcing project leaders tend to disappear before the midterm effects of their work become apparent. Often, the outsourcing will start with a simple *proof of concept* (POC) project that is truly successful. The fun begins when the more complex business systems are moved across. Further, the bulk of the costs of any outsourcing project may not be apparent until a year or two after the outsourcing is completed. A friend of mine who works in sales mentioned how the outsourced system at his company meant that he now wasted much more time reporting and explaining issues than before. This lost time is hard to measure and remains unaccounted for.

The more levels you have between the stakeholders and workers, the more can get lost in translation. When teams no longer share the same office, the communication paths become longer and more error-prone. Modern work methodologies rely on close collaboration within the project, but outsourcing to an external company can make this collaboration much harder to achieve. Responding to change can also be more difficult. Getting across the exact requirements is already a huge challenge, and communication can be difficult even when the business and IT people share the same native language. Adding further language and cultural differences between teams and external companies only serves to compound the difficulty. Counteracting this problem requires intermediaries, making the chain even longer. Unsurprisingly, many companies have stopped outsourcing the development of their core business software and brought it back internally.

Do you really want the mission-critical knowledge going out of your hands?

There are sometimes valid reasons for outsourcing though, such as a local shortage of IT specialists. And to be fair, outsourcing can be successful for certain tasks. However, the final cost savings are often not as spectacular as expected.

In one case, I saw how a large outsourcing provider took all their best programmers off a project shortly after the go-live and reassigned them to other clients. This left my client's project with a much weaker team to try and fix all the bugs.

Are the people advising the management on outsourcing thinking in the short or long term?

Rarely are any serious investors looking for a quick buck. They generally prefer a sustained economic performance instead. Although there are many smart people and skilled programmers all over the world, outsourcing work to people unfamiliar with the core business processes doesn't make much sense to me.

A major incident in the IT systems at British Airways serves as a cautionary tale. Thousands of pieces of luggage went missing, and many flights were delayed or even canceled as a result. The estimated cost of this issue is in the tens of millions of pounds. Reports suggest that many skilled workers had been laid off not long before and the IT work had been outsourced to reduce costs [British Airways Flight Chaos 2017].

Returning to the book *The Phoenix Project*, the CEO (Steve) to the VP of IT Operations (Bill) says, "There are two things I've learned in the last month. One is that IT matters. IT is not just a department that I can delegate away. IT is smack in the middle of every major company effort we have and is critical to almost every aspect of daily operations" [Kim, Behr & Spafford 2013].

At the same time, you don't always need to bring in more experts for every new IT challenge that crops up—it's good practice to try and fully utilize the resources you already have first. That way you can improve processes and find the root causes of any problems within your company.

Outsourcing may only improve the bottom line in the short term. During most business cycles, your investments in technology should therefore add business value rather than simply cut costs.

Accountability

Many companies need to adopt a culture of more accountability and break the habit of pinpointing blame. These two practices are distinctly different. When no one takes responsibility for their actions, the void that results is usually filled with blame.

How can you take risks, learn from failures, and avoid a blame culture at the same time? Companies and their teams will benefit by speaking openly about their own failures. As mentioned earlier, the architects, project leads, and developers involved in software projects need to be accountable for the burden of maintenance they cause. By upholding a feedback loop, they can learn from any errors. The resulting sense of responsibility will give these team members the incentive to keep future maintenance work to a minimum.

Aligning the Strategy

Some years ago at a company-wide IT department event, we watched a video in which six different business area heads gave a talk. They were all responding to the question, "What do you expect from IT?" The answers from these high-level managers, just two levels below the CEO, were rather revealing. Their expectations ranged from cutting IT costs dramatically to requesting more from IT teams for the same budget, all the way to investing more in IT. Clearly, none of the managers had reviewed the final video before showing it to the 600 people attending. The managers' radically different responses were a complete embarrassment. The audience soon began to chuckle at every new opinion expressed as they were all so contradictory. The hilarious questions from the audience afterward only made matters worse. It looked like the management didn't have a clue about where the company was going and that their IT strategy was nonexistent. Executives need to have a clear picture of how IT will be used to leverage their business, and they need to get the message across in a unified voice. The IT teams can only deliver when there is a definite IT strategy in place.

A golden opportunity for enhancing team alignment presents itself in the common case of a merger or acquisition. Many IT systems will be duplicated, giving each side the chance to benefit from the knowledge of the other. Unfortunately, I've often seen these projects being led according to political decisions instead of unbiased, collaborative ones.

At the same time, it's very interesting to see how some firms have slowly and silently turned into software companies. A senior manager who works for one of the world's largest banks recently told me, "We are actually an IT company, but no one at executive level talks about this. More than 99.8 percent of all our bank transactions are done digitally now."

The Benefits of Agility

Research done by the Boston Consulting Group indicates that agile companies are more profitable and have higher margins than others [Daemon 2017]. In fact, various providers of training courses now teach companies how to achieve overall agility. The focus goes beyond the level of individual projects and addresses the organization as a whole. One of the biggest difficulties companies have though is in adopting an agile culture across *all* departments. Without the full commitment and involvement of business teams, firms cannot achieve optimal agility.

Change is, after all, an inevitable feature of our lives. Learning to be adaptable and flexible in the face of change should really form part of our education, I feel.

In *The Phoenix Project*, Erik (the board candidate) to Bill Palmer says

> *I've long believed that to effectively manage IT is not only a critical competency but a significant predictor of company performance. One of these days, I'd like to create a hedge fund that invests in companies, taking long positions on companies with great IT organizations that help the business win, and short the companies where IT lets everyone down. [...] What better way is there to force the next generation of CEOs to give a shit about IT?*
>
> –[Kim, Behr & Spafford 2013]

To wrap up this section that addresses executives, I'd like to recommend the following three books. You'll find extracts from these in this section and elsewhere in this book. These books will give you many valuable insights into the challenges that lie ahead for you and your organization:

- *The Phoenix Project*

- *Why Simple Wins*

- *The Hard Thing About Hard Things*

Next, we'll consider the preliminary steps you'll need to take at the start of a project.

Setting Out

I'd like to suggest you plan the journey well before starting out. Make sure you take the best path to reach the right peak, nurturing the necessary team spirit and acquiring all the right climbing skills.

The first step is to define all the elements necessary for project success. Then, throughout the ascent, be sure to keep your sights fixed on the overall business objectives ahead.

Project Design

Project design occurs at an early stage of the project when the key features, duration, cost, criteria for success, and scheduling resources are determined. The critical path of the project and risks shall be laid out. The aim is to narrow down the countless possibilities on how to complete a project to a few good project design options that can be used to achieve the desired project goals. Stakeholders can then choose the best design to implement the project. The planning must include understanding the dependencies within the project itself and to outside factors. The project design phase can produce a variety of deliverables, such as diagrams, flowcharts, sitemaps, and more.

Project Success

What are the most important factors for achieving project success? The following list is for large, business-critical software projects. For smaller projects, a subset of these goals may be sufficient:

- **Strategy definition**

  Specify the mission statement or problem definition.

  Define precise overall objectives.

  Decide on buy or build strategy.

  Choose the most suitable project methodology.

  Perform impact analysis.

- **Stakeholder management**

  Identify all stakeholders including end users.

  Consult all stakeholders and align their requirements.

  Assemble the right project team.

  Ensure the project has executive support.

What is your primary goal? It should always be a business goal, and achieving it may require some preliminary technical cleanup work. A stable and easily extendable system may well be instrumental in adding business value. Defining that value, which is usually specific to a project or product, should be a team effort.

Steve Jobs once explained how Apple approached product development quite differently from other companies. Instead of simply utilizing the amazing technology they have, the company first asks, "What incredible benefits can we give to the customer? Where can we take the customer?"

Milestones

The project team should first work on a general road map of where the software should be going. This road map will remain a work in progress as the requirements, goals, and planned work begin to take shape. Project milestones should have no fixed timescale at this early stage as many upcoming decisions may completely change the project plan. Nevertheless, creating measurable milestones is important because vague statements like "The code is 80 percent complete" are hard to measure. A project milestone usually contains one or more deliverables. Your project could include some of the common following milestones, many of which can run in parallel. A few of these deliverables extend over the whole project, whereas the majority are best planned as smaller tasks for each product feature:

- Define problems and objectives.
- Define the initial project language glossary.
- Create the project design.
- Define requirements, use cases, and/or user stories.
- Define the software architecture.
- Define test cases and acceptance criteria.
- Analyze and document impact.
- Deliver on external commitments.
- Build working software—the minimum viable product (MVP)—ready for release.
- Create end user training materials.
- Get user acceptance sign-off.

Project Organization

Your initial project organization will involve setting up a structure comprised of key individuals who can make the best decision on whether to buy or build (see the section "Buy vs. Build" in Chapter 13). Depending upon the size of the project, the stakeholders here may include the project manager, project sponsor, IT manager, application architect, and development lead as well as a business representative and business analyst. Together you need to establish the high-level business requirements—this information will help in evaluating whether any standard software can meet your requirements. As the title of this book suggests, the emphasis here is on software projects, so the build strategy is the main focus. Nonetheless, I strongly recommend you carefully evaluate whether an existing piece of software will fulfill your needs.

To keep the project organization as lean as possible, try to minimize the number of key individuals at this stage, but without neglecting any key stakeholders. It's crucial to get the right balance between too few and too many stakeholders. The core team can always consult specialists from other areas if necessary. I've often seen examples of how "too many cooks spoil the broth" at project meetings where six people represented a single, medium-sized application.

"Organizations must be designed around the people available; not the people fitted into pure-theory organizations" [Brooks 1995].

Responsibilities

Defining the project responsibilities is a task that should be done early in the project and adapted according to the dynamics of the project as it progresses. Once the initial project organization is decided, you can allocate the various responsibilities. Some of these responsibilities may only be ad interim, until all the team members have started on the project. Above all, it's important to keep things flexible.

Imagine that a system has been around in your company for 12 years. Everyone who made the original project decisions has now moved on. You may find an odd *business rule* such as this: "If a customer is resident in Canada, then a mandatory sentence to this effect is required by law on the customer invoice." The current development team can confirm the logic by analyzing the code, but finding out whether this sentence is still legally relevant is the responsibility of the business team.

On the other hand, the IT team is responsible for selecting a software component and for keeping it up to date. Sometimes, the distinctions between responsibilities are not handled well though. Just because the IT team oversees the code doesn't mean they're responsible for the business decisions involved. I've experienced business teams believing that the developer should have noticed an obvious error in the business requirements. Sometimes the business wants exotic things, which the developer may or may not question. A good developer with some domain knowledge will, however, question the code if something appears unusual—they will reconfirm the desired change with the business departments if necessary. The business department defines its exact needs in sharp business requirements, and all the business rules are ultimately the responsibility of the corresponding business department, both during the project and beyond.

The business department is primarily responsible for defining its exact needs. Also, the final decision for going live with a feature lies with the responsible business department.

Methodology

Once the responsibilities are clear, you'll be ready to face the challenge of choosing the right software project methodology. Which methodology is right for your project?

In basic terms, the software project methodology refers to the structure of the project management and the handling of the whole software development lifecycle. The steps in this lifecycle include analysis, design, development, testing, and others. We've already taken a brief look at the two main methodology types—Waterfall and Agile—earlier in the book. Although there are other types and flavors, I'll continue to focus on these two most commonly used ones. The following images help illustrate these two methodologies.

As the next image shows, the traditional Waterfall methodology is *sequential* (Figure 12-3). This plan-driven method involves separating the project into a distinct sequence of steps that include analysis, design, construction, testing, production, and maintenance. The software goes live after a longer period of development that can vary from a few months to 1 or even 2 years. The success of the project can be easily measured if everything comes out according to the original plan.

Figure 12-3. *Waterfall methodology*

Some types of products, like the software for a pacemaker or for traffic lights, typically need to be built in full. The MVP is the final product. A project designed to upgrade pacemakers of a certain model to a newer software version, for example, would be difficult to organize, and any bugs could even put people's lives at risk. A Waterfall approach may work best for this type of project, where the requirements remain stable throughout the project.

Having said that, even Elon Musk's SpaceX company now works with Agile approaches. Therefore, industries such as defense, space travel, and medical device supplies that were associated more with a sequential methodology are now adopting leaner models.

The Internet of Things (IoT) is also changing the way devices are updated. Appliances, such as washing machines and ovens, which were never or only rarely updated, are now also being connected to the Internet.

The rigid structure of the Waterfall, which was actually modeled on the assembly line, is not very efficient when it comes to dealing with changes during the project.

Therefore, as changes are frequently made to the requirements in most types of project, the more flexible Agile approach was initiated as a kind of counterpart to the Waterfall method.

The term *Lean* (Figure 12-4) comes from *lean manufacturing*, which aims to eliminate anything that does not add value. It also means working only on what is needed at any given time. Accordingly, an *Agile/iterative* approach breaks the software project down into smaller parts. Each part contains at least one full feature that can be specified and developed and can go live independently. Currently, the dominant Agile software methodology is *Scrum*, which is a framework for organizing the people, the product, the communication, and the work, using time-boxed sprints. *Kanban*, on the other hand, is a management method that focuses primarily on leveraging work in progress (WIP) limits.

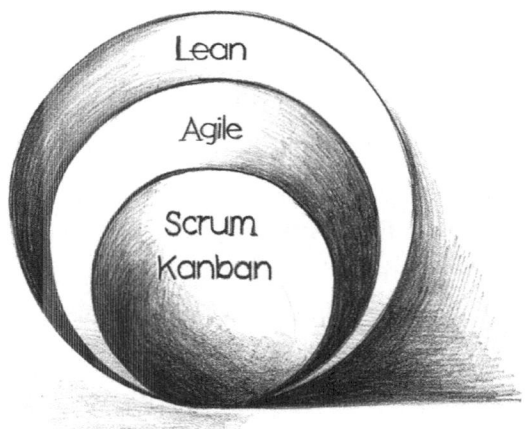

Figure 12-4. *Breakdown of the Lean methodology in software development*

Agile assumes that the requirements will be unstable. The second principle in the Manifesto for Agile Software Development states, "Welcome changing requirements, even late in development. Agile processes harness change for the customer's competitive advantage" [Manifesto Agile Software Development 2001]. I strongly recommend you read the 12 principles of the Manifesto in "Appendix A: Collaboration." The Agile features act as mirrors of how you're working, and knowing how to react to them is crucial.

When selecting a methodology, one option is to let the project teams choose the one that works best for them. A company isn't limited to using only one methodology—different types can be used side by side. I've worked for companies that successfully had Agile and Waterfall projects running simultaneously.

"Iterative approaches are usually a better option for many reasons, but an iterative approach that ignores prerequisites can end up costing significantly more than a sequential project that pays close attention to prerequisites" [McConnell 2004].

The Agile structure ensures regular communication and feedback between the business and IT sides, although it's possible to regularly show the software to the business using any methodology type. However, the two-way communication that is a characteristic of Agile projects means the business teams are actively involved—there will be countless details to discuss and decide on for every project. *And no one should ever feel afraid of speaking up!* I've never had the feeling that anyone asked too many questions during a project—but I have seen many issues crop up due to a lack of communication and the false assumptions that result. To fully reap all the benefits of Agile, everyone, including the business teams, needs to commit to this methodology.

Measuring the success of Agile projects based on the original plan is generally more difficult than with Waterfall because the requirements usually change considerably throughout the project. Of course, these changes will impact the timeline and the budget too. Therefore, the project

111

success should be measured by the business value added. Many traditional companies struggle with this switch of mentality, as they are used to assigning a project budget and trying to stay on time and in budget. Agile is more of an ongoing product commitment.

There are various other benefits to working in an Agile way too. One simplified metaphor of this methodology is that you first build a bicycle, then a motorbike, and then a car, providing the customer with a working product for each vehicle along the way. That way you reach the final goal gradually. Maybe a bicycle is all they need. If the requirements are very clear though, I personally see a cost advantage in going straight for the car.

How can you use Agile methodologies if no one in your teams has any practical experience of this work? I suggest you consult a specialist coach for the first project at least. Just reading the theory is not enough. As a matter of fact, in many projects I've worked on, the requirements could have been specified more precisely in advance. Therefore, using the iterative approach sometimes felt a bit like fighting the symptom rather than addressing the root cause—the team members, fully aware that they could change things later, did not trouble with detailed precision. It costs more to build something and then change it, however, than to build it correctly in the first place.

"Neither Agile nor DevOps are business goals in and of themselves. Both are cultural movements that can inspire your organization with better means for achieving your goals" [Buchanan].

The software project methodology you use is simply a means of getting things done—the business objective is not to chase the latest methodology. There's no guarantee that a product will meet the targets and succeed. Success depends instead on delivering a product or feature that adds business value and achieves the main objectives. Having a supporting framework for working together will assist in achieving this goal.

I've experienced a project where Scrum did not work because the product owner had no understanding of the technical side. They simply placed all the business topics, which they understood very well, at the top of the priority list. Although the team claimed to use Scrum, it was effectively still working according to an old-fashioned hierarchical structure. You need to focus on *collaboration* between all team members instead. Just because a project labels itself Agile doesn't mean it really is Agile. The point is that everyone really needs to understand and follow the principles. That way, every opinion counts and trust builds up, resulting in teams that function well through greater autonomy.

Managing Risks

Every project comes with its own inherent risks, and teams should never shy away from discussing and addressing these risks. It's crucial that the risks, such as budget overruns, project cancellations, security risks, and key persons leaving, are all managed well. As the book *Peopleware* states, no risk should ever be classified as "just too awful to think about" [DeMarco & Lister 2013]. The project participants and stakeholders can contribute to the task of identifying risks, which should then be classified and prioritized. Teams can work on risk mitigation to reduce or even eliminate these risks. As mentioned in Part 1, "Conceptual Guide," if you've opted for a bold experimental approach, you can expect higher costs and project delays as well as higher risks.

Now I'd like to look at the crucial aspect of communication and collaboration both within and between teams. Drawing on my own experience as well as that of others, I'll offer some practical advice on how to get everybody rowing the boat together.

Interacting Successfully

This section of the book is one of the most important, I believe. The reason is that most delays, budget overruns, and project failures are not due to technical reasons. The fault usually lies instead in poor communication and inadequate collaboration both within and between the various teams involved. The larger a project is, the more of a challenge aligning the information becomes.

Good two-way communication brings mutual understanding, motivated team members, clear objectives, and a fun project to work on.

Effective communication is certainly important in all types of projects, yet I believe it is particularly crucial in software projects. The reasons are as follows:

- Close collaboration between IT and business teams is essential.

- Not all IT specialists are keen communicators.

- The IT teams need to understand the business requirements in depth.

- Hundreds if not thousands of decisions are necessary.

- Different fields of work combine—each with their own jargon.

To use an example from construction, the home owners will probably communicate mostly with the architect and may never need to speak to the individual bricklayer. This is not the case with software projects though. In practice, the business stakeholder and the development team have many fine-grained decisions to discuss, such as clarifying the requirements, demonstrating the user interface, discussing a bug, and so on.

Active Listening

Good communication starts with listening. You might think this is stating the obvious—but in my experience, a lack of *active* listening is one of the biggest causes of errors in software projects. As listening is often the missing half of communication, we should look at this aspect before moving on to two-way communication. Whenever I catch myself losing attention when I should be listening, I ask the person to repeat themselves where necessary. In my opinion, even some very smart people lose a lot of credit by being poor listeners. At the same time, some of those who consider themselves good listeners often aren't, or they can only listen well for a very short time.

As a rule of thumb, the speaker or sender should first assume that their message has not been received in the way it was intended. By the same token, the receiver should assume they have not understood the message in the way it was meant. Important decisions should be briefly reconfirmed, ideally in written form. A sign of a good listener is patience. Some listeners make assumptions and think, "Oh, I've heard this all before, and I know where it's going." Consequently, they miss the main part of the message, which could be quite different from what they expected.

Katie Owens, in her TEDxYouth video, succinctly describes the essence of active listening in three short words: *be here now*. She also specifies the way to achieve this receptive state of mind [Owens 2013]:

- Stop being in a rush and tell yourself you need to live in the moment. That way you really will *listen* to the person speaking.

- Physically distance yourself from distractions by putting your phone out of reach and moving away from the computer screen.

- Force yourself to stop thinking about your plans for the evening and focus on *being present.*

- Let the other person really see and feel they have your full attention. Making eye contact without ever staring is vital.

- Whenever they make an important point, interject with something like "So what you're saying is..." This gives the other person the opportunity to say, "No, I actually meant..."

- You can then rephrase their words and ask for confirmation that you've understood the point correctly.

- After the speaker has finished, he or she will feel satisfied that you've listened and understood properly.

Once you start practicing such active listening, you'll be ready for better two-way communication.

"There's one technique that you must use if you want people to listen to you: listen to them" [Hunt & Thomas 1999].

Adopting active and effective listening in your project and in your company is a very good start. Give everyone your undivided attention. Let people finish what they're saying and allow them to get their point across. Avoid interrupting and cutting people short. Pause sometimes when you're speaking yourself, to let people ask questions.

Effective Communication

Now that we've examined the listening part of interaction, we can move on to the actual communication. The business and IT people in your teams will probably come from different professional backgrounds and think in different contexts. By establishing common ground and good communication both within and between teams, you'll greatly enhance the project outcome. There's no doubt that effective communication lies at the core of every successful software project.

Assumptions

Misunderstandings between people happen all the time, and I'm sure you've experienced your share. One of the main causes, I believe, is false assumptions. This was a topic I thoroughly enjoyed reading about in Prof. Steve Peters's book *The Chimp Paradox*. He explains how we use assumptions and instincts as a form of survival to protect ourselves from harm. These methods were very useful in the jungle, when we had to make rapid decisions on whether to run, fight, or freeze—but such basic instincts are obviously ill-suited to our more complex modern-day society or to business contexts [Peters 2013].

Many business decisions require careful consideration and an understanding of a variety of factors. The more light we can shine on the blind spots, the better we get at making rational decisions. If you don't have all the information necessary for a decision, it's always best to ask.

Eliminating assumptions and replacing them with facts, figures, and effective communication should be a major goal in every software project.

Excessive Jargon

The use of jargon is another barrier in communication. At a memorable presentation I attended a while ago, a manager from the business side was addressing an audience of IT employees. They began by saying, "As I don't have an IT background, I'm often overwhelmed by the excessive jargon used at the presentations given by IT staff. Therefore, I'll first give you a brief introduction into the core business terminology you need to understand my presentation." I found this approach incredibly powerful, and it also went down well with the rest of the audience. This kind of introduction should be the standard when not all attendees have the same background. You'll find that few people will openly express their lack of understanding otherwise, particularly at a presentation or large meeting.

Who at the meeting or presentation has what level of knowledge? Do the attendees have sufficient technical or business understanding to grasp most of the content being presented?

One of my early professional experiences was presenting a new feature to an audience of about 12. Company policy specified that a representative from each relevant department should be present. Most of these were non-IT people. I felt sure the main discussion topic would be the slow network connection in the company stores. We had spent a considerable amount of time improving the system performance, and I had come well prepared to discuss how much data should be sent across the line. To my surprise, however, we spent most of the time discussing the text label on one of the buttons, and the performance issue was sidelined. This experience was a real eye-opener for me. In retrospect, I can understand it—the text on the button was a safe topic that everyone could comprehend. The network issue was obviously more complex and involved a fair amount of jargon. Had I known this at the time, I'd have explained the technical terms to demystify the performance issue, ensuring that everyone in the room could follow. The more minds that consider a topic, the better.

I've even worked with experienced IT professionals who used excessive jargon. They may do this to show off their superior knowledge, to appear profound and/or to get their suggestion pushed through. Sometimes their strategy works. Nonetheless, they largely fail to achieve the goal of communicating well with others. If you encounter such a situation, don't be afraid to say, "Please, can you explain this in simple words?" For me, this willingness and ability to describe a complex IT topic in words that non-IT people can follow makes a truly great IT professional.

Ubiquitous Language

Another element teams use to achieve good communication among everyone in a project is the *ubiquitous language* specific to the project. Coined by software expert Eric Evans, the term refers to the project language that teams adopt to avoid ambiguity. It describes the distinct set of precise terms to be used in this particular business domain. Project members work together to define these business-specific terms. The language is used to discuss the domain, to define precise requirements, and finally it's used in the code itself. Any new team members will first need to learn the ubiquitous language—everyone on the project needs to understand exactly what's meant by each term. In other words, this language goes a long way in preventing misunderstandings.

The word *net*, for example, can have different meanings depending on the context. In finance, the *net income* refers to the amount that remains after deductions. It can mean the final outcome, the net result. A further usage of this word is for an entrapping device or the net used in various sports. In technology, the word is an abbreviation for the Internet. The word has yet other meanings too. Many terms used in business can be ambiguous too.

As much as I support the idea of using such a language, I'm not a big fan of the term "ubiquitous." In my experience, many team members have struggled with it. As this project language is central to a project, I'd prefer a simpler and more descriptive term. "Specific domain language" or simply "our project language," for example, would work well.

Differences of Opinion

People will always disagree over certain topics within a project. Allow for such disagreements, but never let them get personal. Stick to a factual and objective approach when speaking. Use examples, figures, and statistics to support your point. Let the group decide democratically which path should be followed, and let the ones who disagree express their objections too. Some people have trouble if things don't go their way though and start to be systematically unfair to their colleagues. This kind of behavior should never be tolerated. Instead, allow for the right to consult an independent referee if someone is absolutely convinced the group has made the wrong decision. Let's not forget there are cases where a crazy idea later becomes the cash cow of the company.

Frederick P. Brooks Jr., describing the vibrant discussions at the faculty meeting of the computer science department (at the University of North Carolina (UNC) at Chapel Hill), says, "We are very good at courteous disagreement. I have never seen people leave a faculty meeting mad at each other over the arguments. I've seen them disagree very vigorously— *good*—we want to speak honestly. We want to express opinions precisely, but courteously" [UNC Computer Science 2015].

During a conversation, especially a heated one, most people formulate their response before the other person has even finished their statement. This form of communication is more verbal combat than exchange of ideas or opinions. Avoid this reflex by slowing down. Rather than rushing to reply, take a moment to consider the other person's statement. Ask follow-up questions to better understand what the speaker intended. Try to understand the deeper motivations behind their statement. Remember you don't need to share someone's opinion in order to understand it and acknowledge it and listening will help inform and expand your own opinion.

–[Fowler Chad 2014]

Eventually, a team needs to reach the point where they commit to a decision after any initial disagreements. Further analysis or discussion may still be required for critical decisions. It's important to remember throughout that a successful project depends on the commitment of all group members to one approach.

Dos and Don'ts

Here are a few ideas to help make the interactions between team members go more smoothly. If these suggestions aren't right for you, you'll benefit from compiling your own list of dos and don'ts. The most important message here is for you and your team to decide on the best modes of communication, adapting these throughout the project according to your needs.

Do

- Introduce one another.

- Let everyone voice their opinion.

- Listen to each other.

- Try to communicate using dialogues instead of lengthy monologues.

- Send customized information to team members where appropriate.

- Collect questions to ask in one go rather than disturbing someone every 20 minutes.

- Give credit where it's due.

- Stand in line and wait for your turn, unless it really is an emergency (it usually isn't).

- Foster good decision-making practices.

- Send links to documents rather than huge attachments where possible.

Don't

- Talk in a quiet monotonous voice—this makes listening a real strain.

- Finger-point.

- Broadcast information to everyone that's relevant only to a few.

- Overload slides.

In a nutshell, "the fundamental answer is thoroughgoing, careful, and sympathetic communication" [Brooks 1995 p. 54].

Feedback

Honest feedback is great for keeping everyone in the picture and helping them stay on track.

Instead of conducting annual reviews, companies will find it pays to adopt a culture of continuous feedback.

This regular feedback should relate to the work that is currently underway; it should be constructive and should build up self-awareness in the team members. The primary goal of this type of feedback is to build on people's strengths. You'll find it worthwhile to regularly touch base with every team member individually throughout a project to see how they're progressing. Be sure to adopt a positive tone when giving feedback, and take care to make any criticisms in a constructive manner.

According to Ben Horowitz, feedback is a dialogue, not a monologue:

You may be the CEO and you may be telling someone about something that you don't like or disagree with, but that doesn't mean you're right. Your employee should know more about her function than you. She should have more data than you. You may be wrong. As a result, your goal should be for your feedback to open rather than close the discussion. Encourage people to challenge your judgment and argue the point to conclusion.

–[Horowitz 2014]

As well as giving feedback, a team leader or executive can also benefit the project by collecting feedback—a team survey is a good way of doing this. In one large project I was involved in, all the team members participated in an anonymous survey about one-third of the way through. The goal was to check how the project members felt on questions such as reaching targets, project administration overheads, communication, job satisfaction, and so forth. I can highly recommend conducting such a survey as the feedback can be very helpful.

Motivation and Demotivation

Extrinsic and intrinsic forms of motivation are quite different. The extrinsic kind relies on external rewards such as financial benefits, whereas intrinsic motivation comes from within the individual. Companies and teams that focus on and encourage intrinsic motivation benefit the most. People generally do have an intrinsic motivation to learn and achieve a good result. If a team member's only motivation for making their way to work every day is the salary, this will not be sustainable. Give them a definite purpose, let them know their work is important, and you'll solve the problem. One effective motivator already mentioned is to give each team member detailed, personal feedback after keeping track of their work. It might take a bit of time, but you'll see a high level of intrinsic motivation in many of the project members as a result. Conversely, the more this intrinsic motivation is neglected, the higher the demands and expectations will be for extrinsic motivation, which is less satisfactory all round.

Creating an atmosphere of openness and tolerance also encourages people to contribute more wholeheartedly. Amy Edmondson, who is well known for her work on teams, makes some very interesting points on this topic. Using her research from a variety of industries, she shows organizations how to change their management style for the better. By leaving behind the "name and blame" game and adopting an attitude that *embraces* failure instead, everyone can learn important lessons for the future. Amy uses the term *psychological safety* to describe this practice and shows just how much companies can benefit from it. She defines the term as "a belief that one will not be punished or humiliated for speaking up with ideas, questions, concerns or mistakes." The first step when introducing the new atmosphere of openness is to "acknowledge your own fallibility," she says. You can then start encouraging people to give voice to their natural curiosity and ask plenty of questions.

Once employees feel confident that their contribution will be valued, it will pave the way for increased intrinsic motivation. Amy is keen to add that granting psychological safety does not affect accountability, which she sees as a separate dimension. While accountability has its place, she says, "it's equally important to free people up [to] really engage and not be afraid of each other" [Edmondson 2014].

In one instance I know of, an IT team had made a big mistake, and the head of IT addressed them at a huge IT department meeting. They embarrassed the team members by asking them to avoid making such mistakes in the future. Well, we all make mistakes, and nobody wants to be named in front of everyone like that. This kind of reaction to errors simply results in people starting to hide mistakes, thereby making it very hard to find the root cause. Time is wasted and a blame culture introduced. Using a different approach, a team at another company introduced a biweekly failure meeting. Everyone had to present three mistakes they'd made or describe an area they were struggling with in their work. The meeting was not only fun, but everyone learned from their own and others' mistakes and often found solutions collectively.

How open is your company when it comes to making mistakes?

Mistakes are one of the final taboos to overcome in a professional environment. We can discuss almost anything these days, the work-life balance or the pros and cons of being available for 24 hours a day, but we prefer not to mention our mistakes. After all, without an open culture, admitting we've made mistakes could weaken our position and even damage our career.

In the section "Fail Fast," I'll discuss how Spotify embraces a continuous improvement culture that includes learning from failures.

Nothing unifies people better than complaining about IT. She has an uncanny knack for blaming other people for her screw-ups, especially IT people.

—Gene Kim, Kevin Behr, and George Spafford [Kim, Behr & Spafford 2013]

Praise in public; criticize in private. I'm astounded at how some people just ignore this familiar principle. Being criticized in front of others is demotivating for anyone, and it makes the rest of the group feel awkward too.

Generally, demeaning remarks regarding whole groups of people only serve to foster resentment. Comments such as "Why can't the IT people ever get it right?" or "The business just doesn't know what they want" are not only incorrect but also counterproductive. Making such sweeping generalizations does nothing toward resolving any current issues. Stereotyping an entire profession or even a whole family of professions is

125

unfair. Such generalizations can become toxic and result in an unwanted blame culture too. Project members need to cultivate a culture of collaboration between everyone and avoid the *us-vs.-them* mentality that can so easily arise.

Differences of Opinion

Even when speaking to capable IT professionals, I've sometimes been left feeling that the only way to climb the mountain is their way. IT professionals with such a single-solution attitude will defend their position endlessly in any discussion. I'm sure this happens in other professions too. Individuals do sometimes like to differentiate themselves from their peers. But how would you feel if your doctors heatedly discussed the best treatment for you in your presence? Witnessing this as a patient, you'd probably doubt their abilities and wonder who was right. Thankfully, such differences of opinion are not usually expressed in front of a patient. I've been in IT project meetings, however, where two or more techies had argued about technical details in front of business colleagues. Although these passionate discussions are necessary, the extensive use of technical jargon completely confuses non-techies and wastes their time. If you ever find yourself trapped in this kind of situation, just ask the IT people to resolve the matter among themselves. Once they've settled things, the team members can all reconvene.

Managing Stakeholders

Involving and managing all the stakeholders is something that must not be overlooked. The stakeholders include all the project members, subject matter experts, and sponsors as well as any surrounding systems' teams.

The stakeholders play the biggest role in defining the requirements of the new system. If the individual stakeholders' goals are partly contradictory, aligning these requirements is a critical step in defining the final solution. Managing stakeholders effectively is one of a project manager's main tasks. As there's plenty of information on this topic within project management literature, I won't go into any great detail here.

Three potential risks should be mentioned though. First, you should ask: Is there anything that poses a threat to the project and a financial risk to the company? Is there anyone who has a personal interest in seeing the project fail? It's very hard to prove that someone wants a project to fail, as they are unlikely to admit it openly. They may even join forces with a few of their trusted work colleagues and quietly try to obstruct a project. A desire to sabotage the project may stem from their preference for an alternative solution, or they may feel the project will adversely affect their career. The project leader should carefully consider this matter and try to get everyone's goals aligned, involving the upper management and even the board if necessary. It's vital to ensure that no such personal interests can seriously endanger a project outcome.

The second point concerns any job losses that may come because of the project. Will the new software automate the job of an individual or a team? This question may or may not be closely connected to the first point. An automation project that frees up the business department from a manual task should be embraced. Instead of spending many hours or days each month on such laborious tasks, the teams can focus more on their core competencies. Your company should welcome this automation that can free up resources. Lisa Bodell nicely illustrates this point in her book *Why simple wins*. Describing how the credit union Vancity aimed to improve its members' experience, she says

> *As the executive team saw it, corporate overhauls often floundered when leaders tried to automate too much in the interest to cut costs. Rather, they wanted to upgrade the way the credit union served its members. That meant using technology not to replace staff but to empower them to do more of what they had been doing so well in service to members.*

Lastly, be aware that any important stakeholders who are being neglected may become critical of the project. Be sure to bring in any such stakeholders to actively participate and contribute throughout the lifetime of the project.

Communication Equipment

As good communication is central to the success of a software project, companies should ensure they have the appropriate equipment. Although it seems obvious that a company needs well-equipped meeting rooms, I've often been surprised at how inadequately some rooms were equipped. Video projectors don't always cater for the resolution of the users' devices, for example. Due to such basic inadequacies, the presenter may be unable to show the software correctly. The simple purchase of a short-throw projector, for instance, which can create an 80" screen even in a small meeting room, would solve this problem.

Where the team members are together in one place, using physical boards on the walls for making notes and discussing team topics can be a great way of interacting. Many of the Agile methodologies strongly encourage using handwritten notes on walls and boards rather than digital ones to track work and communicate other information. I can highly recommend this technique, which can, of course, be used in any type of methodology.

"Because work can be assigned to people in more ways than ever, we want to make visible our existing commitments" [Kim, Behr & Spafford 2013].

At this point in the book, I'd like to stand back and consider the philosophies that relate to project management. The insights you will gain here will lay the foundation for the concrete steps that follow.

Software Project Culture

You may think that discussing philosophies in a technical guide is a little out of place, yet as already mentioned, the quality of our collaboration influences the outcome of software projects more than any other factor. Expecting to fix a problem just by throwing more money at the situation is usually ineffective. Instead, developing some awareness of the deeper underlying processes and interactions involved will greatly heighten your chance of project success.

Leadership

Micromanaging team members during software projects is usually quite unnecessary. This management style should only be used for an individual in exceptional cases where the common approach hasn't worked. The primary role of managers and project leaders should be an enabling one. It's their responsibility to oversee and assist the professionals to complete their project tasks efficiently, on time, and to the desired quality. The leaders need to be continuously coordinating the work processes among individuals and teams.

In the book *Project Pain Reliever*, Ty Kiisel advocates the following management style: "Great leaders should avoid the command-and-control type of leadership and move instead towards consensus decision-making; where everyone's best ideas are taken into account" [Garrett et al. 2012].

Knowing how to spot hidden leadership talent is a big advantage for any organization. A recently published article in the *Harvard Business Review* gives some pointers to help you identify potential leaders among your team members. The article shows the results of a study in which the behavior of team members was monitored over a specific period. "The number of 'cheers' an employee received was highly correlated with high-network influence. This means the group of people who received the most praise from colleagues acted as communication hubs for the entire organization and were central to work getting done." It turned out that high-performing employees spent almost 4 hours more every week closely collaborating within the team and had larger internal networks than their peers [Nielsen, Niu & Meng 2016].

Working Together

In this section, we'll touch on some philosophies of working together as a team. Later, in the section "Teams" of Chapter 13, we'll look at some of the more practical details of forming a team. One of the initial steps in every project involves putting together and organizing the team in a way that works best for the project. Whether team members have worked on projects together before or they are newly assembled, it's great when they make a fun and effective team. Working in such a team makes all the difference to job satisfaction, team communication, and the outcome of the project.

The optimal model for team development, according to one expert, involves four stages–*forming, storming, norming*, and *performing* [Tuckman 1965]:

- In the *forming* phase, the newly assembled members meet and get to know each other. They learn about the team goals and gradually get to know their tasks.

- *Storming* can involve criticisms and disagreements such as questioning the leader and attempts to dominate, which all need to be resolved before the next stage is reached.

- During the *norming* stage, team members begin to share a common goal, and a spirit of cooperation emerges.

- Finally, the *performing* phase begins, when team members are competent, can act autonomously, and are able to contribute to the team decision-making process without supervision.

One example of structuring the way two people work together comes from a relative of mine. He had two good programmers in his team who weren't achieving the expected results. As the team leader, he knew that one could only write good code when he wasn't given too many tasks at once. Overloading him tended to make him work a bit chaotically. The other developer liked everything to be structured and tidy and got a bit annoyed when his colleague's work was messy. The problem was solved when my relative made the well-organized guy responsible for passing on the tasks to the other one. The tidy one simply gave tasks in smaller parts sequentially to the other developer, and this arrangement suited them both well. The slightly chaotic guy was no longer distracted and had less on his plate to worry about, and they both delivered higher-quality work in a shorter time. Such small adjustments in a team can turn a frustrating situation into a positive one.

Try to resolve any issues as early as possible. Most issues start small and only get bigger when we turn a blind eye to them. If two people have a disagreement, they should first sit together and try to resolve the difficulty between themselves. If this doesn't work, then it's best to bring in a third person from the next level up. Everyone should be heard without interruption. This kind of open discussion will, in fact, solve most issues. Otherwise, consider getting help in conflict resolution, either internally or externally. Accepting the need for such help and then requesting it is a positive step and a sign of strength. Bring in mediators and team-building specialists where necessary. They can enable individuals or teams to improve the way they work and communicate together. Not every such specialist will be suited to your team—you may need to try different specialists and approaches, but don't give up. If you are aware that major issues need to be resolved within a team, then keep at it. There are many positive examples of dysfunctional teams getting back on track. If you do nothing though, the problem is sure to get worse.

Fail Fast

As we've already seen, business is now beginning to embrace failure as part of the new culture of openness. As such, *failing fast* involves a complete change in mentality in which a project or even a company views failure as a positive learning experience. Companies inevitably need to try different strategies—a failed project could eventually turn out to be a valuable experience that points the company in the right direction. The Swiss company Zuehlke recently tweeted the following motto: "Think big, start small, fail cheap, move fast!"

One of the best learning videos I've seen on the topic of failure is from the digital music provider Spotify. They even publicly announce some of the mistakes they've made and what they've learned as a result. "We aim to make mistakes faster than anyone else," says company founder Daniel Ek. The company takes the view that building something good means making mistakes and that failing also means learning. Spotify uses this strategy of failing quickly, learning quickly, and then improving quickly to achieve long-term success. By giving team members the freedom to boldly experiment without any fear of failure, the company allows for great new ideas to emerge. The attitude toward making mistakes involves an emphasis on *fast failure recovery* rather than *failure avoidance*. Also, because "failing without learning is just failing," Spotify follows up mistakes with a *post-mortem* session. No blame is attached to the failure during this session. Instead, the team aims to discover what went wrong and why and what they should change. In this way, the whole team can expand their knowledge just from one member's mistake. There's also a strong focus on improving the *process*, not just the *product*. At a meeting held once every few weeks, all the teams taking part in the project review how well the process is going. Importantly, this working culture of continuous improvement is "driven from below and supported from above" [Spotify Engineering Culture 2014].

When someone makes a mistake, either they will realize it themselves, or someone else will notice it. In the second case, I firmly believe it's best to inform the person who made the mistake directly—this should be done in private and in a friendly manner—giving the person time to come forward themselves or investigate if the problem has a different root cause. If there is no fear of reprisals, this approach works well most of the time. In instances where errors are made too often, either by a person or a team, you'll need to identify the cause. The work process may be faulty— can the team find a way to stabilize the process? Additional job training may be required, or if the right skills are there, the person may require concentration training. Try to work with the people you have, rather than following the expensive and counterproductive practice of rapid hire and fire.

Having said that, several factors together may be the cause of a problem. Although our inner instincts push us to identify a single person to blame, when we examine the issue more closely, we may find the cause is more complex. Often a chain of smaller problems lead to a bigger issue.

Sharing their mistakes with the public is certainly a new trend for companies. The trend is catching on quickly as conference participants now follow suit, sharing their failed business plans or even personal mistakes. Nonetheless, the norm is still to proudly present success stories and sweep the failures under the carpet. Companies worldwide can all benefit by embracing their failures, however. Why not stop the finger-pointing and blaming to share the lessons learned instead? A culture of error tolerance and open discussion does undoubtedly lead to better results.

Time Management

Good time management is crucial in every project. Sometimes, external commitments or new regulations can dictate the schedule. For example, a website could be part of a wider marketing campaign, where various other deliverables, like a TV spot or billboards, would all need to be ready at the same time. Delays in the software project can hold up everything else and cause high costs for the necessary postponements.

Further, if the early phases of problem definition and requirements analysis are delayed, teams are often expected to catch up later down the line. This is a very bad solution, and I haven't seen it work once. As a developer, I find nothing more frustrating than being landed in this kind of situation. If the definition of requirements took longer than expected, it was probably due to the requirements being more complex than expected. To still meet the deadline in such a case, the scope must be reduced if the minimum viable product (MVP) allows it. Alternatively, if the scope is retained, the deadline must be set back. If no adjustment is made, the quality of the software will suffer. The best way to proceed through a project is therefore to regularly adapt the plan according to the progress.

Today more than ever, by releasing software prematurely and disappointing users, you'll find it hard to recover from a bad rating.

Make sure you avoid pressuring people or teams to get things done ASAP when your team won't be ready to use the software for a while. In one such situation, our team was placed under pressure, so we worked weekends to get everything ready. The discovery later on that the rush had been for nothing left us feeling pretty devastated. As a result, the business may find it difficult to get the team motivated for a genuine deadline in the future.

Business Logic

Basic functionality, such as sending emails or storing data, is the same across many different systems. The *business logic*, however, comprises the specifics of an industry, the laws in the countries your company operates in, internal company rules, and so forth. The term *business logic* also represents the real-world business rules in software systems. These rules, policies, and workflows are an essential part of software projects. Examples of business rules would be the conditions of a customer contract or a rule that a customer may purchase only one item of a special offer product.

The business rules can be varied and extensive—identifying and defining these rules is a fundamental prerequisite in every software project.

I find it very useful to differentiate between what I call *soft* and *hard* business logic. A *soft business rule* could apply to an address form for customers to complete on your website. One of the fields is the country the customer lives in. A common feature in such a country list is for the most common customer countries to be at the top—this makes it easier for most customers to find theirs (Figure 12-5). Below these country names, all the other countries are listed in alphabetical order. To me this is a soft, or nonessential, business rule because the system would still work if the top six countries were just part of the alphabetical list.

Figure 12-5. *Listing your customers' most commonly chosen countries first can be categorized as a soft business rule*

On the other hand, the list of countries showing store locations is a *hard business rule*. The list must contain all the locations of specific physical buildings. Additionally, local laws may require that an official company address is provided.

If a project is pressed to hit a deadline or reduce project costs, these definitions can be incredibly helpful. The hard and soft business logic can help you determine what can be left out for now.

Thinking Things Through

In nature, a tiger will eat its prey and just walk on once the hunger is stilled, leaving the leftovers for other animals or insects—there's no need to clean up afterward. To some extent we humans act the same. However, with the invention of materials that will last for thousands of years or longer, we face new challenges.

Everyone knows the saying, "It's not rocket science!" Despite their fascinating achievements, however, the rocket scientists are solely responsible for the thousands of tons of waste rotating around the earth that the rockets and satellites have left behind. We're very clever at inventing things and adding more, but as individuals and even as a society, we don't yet place enough value on cleaning up afterward or on putting money aside in advance for any cleanups later. Software projects are not just about adding new features or attracting new users. For me, a successful project is one that's been fully thought through from start to finish, which means clearing up too. Cleaning up unused parts and handling the leavers and movers of the application are intrinsic parts of this work.

The internal movers in a company are one of the most common causes of substantial damage to their organization. In recent years some employees have caused scandals by holding the rights both of requester and approver at the same time. This anomaly typically comes about by someone having the system rights of their previous as well as their present department. Substantial company losses have occurred due to the failure to clean up these user rights. Consequently, most audits these days focus on who has access, how the rights are managed, and how the changes in access rights are tracked.

Documentation

Having accurate product documentation is crucial, but finding the right level of detail can be tricky. Documentation that isn't up to date can be useless when you're trying to find out why a system behaves in a certain way. Outdated documentation creates a dependency on developers to explain how the system works.

Imagine a new employee starting and being given the end user documentation for a system. If the screenshots were outdated, how useful would this learning document be? Detailed, internal documentation can be just as much work to maintain as that for commercial software sold to millions of customers. Therefore, decide in advance what level of detail is right for you, as that sets the bar for all future software releases.

> *By keeping documents minimal and focusing them on complementing code and conversation, documents can stay connected to the project.*
>
> —[Evans 2004]

"Only the code tells the truth" is a common saying in IT. Indeed, it often happens that the only way to know exactly how something works is by consulting a developer and analyzing the code. In supporting various systems over the years, I can confirm that many business enquiries result in having to analyze the code. Enquiries on exceptional cases and complex business logic are especially frequent. But what if the code isn't always entirely correct? This situation can lead to a minor search project to scour through change requests, emails, and meeting minutes.

Therefore, you should make sure that *all* the business decisions are well documented, taking particular care to keep track of the business rules. This documentation acts as the truthful witness of the system and needs to be managed and maintained by the business team responsible. One way of ensuring your documentation is kept up to date is to give documents or web pages a lifespan—set a mandatory expiry or review date at the time of creating the document.

Using change tracking is very useful throughout a project, I find, but it's frequently undermined by people sending out a copy of the same document to a large group of stakeholders without using this feature. Bringing the changes back together is a tedious task.

In most projects, documentation should be kept as lean as possible (Figure 12-6), as described by Alan Ackmann, "Conciseness and efficiency of language can be like backpacking, where an important goal is to carry as little weight as possible" [Ackmann 2016].

Figure 12-6. *Documentation goal: carry as little weight as possible!*

Further, Ackmann describes the importance of writing with clarity. Clear writing avoids ambiguous terms, jargon, heavy phrasing, misleading connotations, incoherent paragraphing, and shaky grammar.

Now that we've looked at software project philosophies, I'd next like to familiarize you with some of the far-reaching decisions and choices you'll need to make at the start of a project.

CHAPTER 13

Define

This chapter focuses on the more specific decisions and steps needed to lead the project to the desired outcome. We start off by looking more closely at one of the earliest questions a software project team should consider, "Shall we build or buy the software?" This decision will strongly influence the further assembly of the team and the internal knowledge required for the software. Our main concern in this chapter will be more with the "build your own" model than for buying and configuring standard software. If your project does decide to buy off-the-shelf packaged software, you'll find that the sections on requirements, testing, and training users are still relevant.

Next, we'll look at how to assemble the right team and how to approach the important area of requirements, based on the methodology you choose. We'll then focus on all the preparation work necessary to ensure you build the software that meets or even exceeds the users' expectations. To wrap up, I'll describe some common pitfalls and suggest ways to prevent or resolve such issues.

Buy vs. Build

As already mentioned in Part 1, "Conceptual Guide," the decision between in-house development and the purchase of packaged software should largely depend on the analysis of the business model and business processes. If the scope of your project is based mainly on industry-standard

© Jonathan Peter Crosby 2023
J. P. Crosby, *The Business Manager's Guide to Software Projects*,
https://doi.org/10.1007/978-1-4842-9231-0_13

business processes, there's a high chance that a range of standard software will be suitable. At the same time, as software development may not be your company's core skill, purchasing standard software might be the simplest option. Where standard software will clearly not meet the company's specific requirements, either an external company can write the software for you, or you can form development teams in-house. Developing software in-house will not only give you full control of the product but will also keep the know-how and skills within the company.

Let's say a start-up company needs to decide on the type of email software to use. They'll probably consider existing email programs for their employees' use. However, if they are an online marketing company, they may also need custom software, as they'll handle the correspondence sent to tens of thousands of customers. Beyond the potential technical limitations of sending so many mails with standard email software, they'll probably need custom software to measure the success of the campaigns. At the same time, as some of today's innovations and niche markets will be tomorrow's norm, the standard software keeps expanding to cater for current trends and reflect the recent innovations.

Choosing standard software before all the major requirements are clear is, in my experience, one of the main reasons for the software failing to meet the client's needs. I've seen many examples where configuring standard software became incredibly complex and ended up taking even longer than building new software. Even after investing considerable time and money, the result was still less than optimal. Having to invest a lot of time configuring standard software probably means that your company has nonstandard business cases. Besides, the underlying performance for standard software may not meet your needs. I've witnessed cases where the provider refused to believe the performance of their software was inadequate for the client's purpose. In one rather complex case, a client of mine spent months with several people working on the analysis, to prove the standard software was causing an issue. When this kind of problem crops up with your own custom-developed software, it can be examined in more depth and adjusted because the code is internally owned. You may find yourself in the common situation of having no control over purchased standard software if the provider doesn't take your issues seriously enough.

Standard off-the-shelf software that has been adapted and customized considerably to meet the customer's needs may no longer benefit from an upgrade to the latest version, as the software may no longer be compatible.

For a different client, I was involved in the replacement project of their invoicing and fee calculation system. The new system needed to handle numerous tailor-made contracts. To my surprise, the client decided to purchase a standard piece of software. The finance department had explained their models and special cases in great detail. The software vendor had a really great product, and their team was highly skilled. Having sold the software to much larger companies, the vendor believed they could easily handle a smaller fish. However, these larger clients had, in fact, mainly used standardized fee contracts. The project subsequently faced delays because the customization took much longer than expected. As the vendor was at risk of making a loss on this project, they tried to charge every special case as a change request, despite their fixed-price deal to customize their software. The software did go live eventually, albeit 2 years late. This example illustrates the careful consideration necessary when deciding whether or not standard software is suitable for your main requirements. Simply put, if you have a bespoke business model, then you probably require bespoke software solutions to get the most out of your software.

There's clearly no easy one-size-fits-all solution when it comes to the question of buy or build. As a software developer, I admit to having a bias toward building. Despite the potentially higher cost, I believe that in many cases, in-house development of the core business systems is best. This solution gives a company all the independence they need for their critical and often unique software. The most successful companies across various industries typically build their own core business systems.

Teams

The benefits of building up an effective team are evident, but it's much easier said than done. Managing a group of exceptionally talented individuals who aren't team players can be like herding cats—each one wants to do their own thing. A team that works well together can work wonders—each member brings complementary strengths and experience to the project. In this section we'll look at how you can achieve the most potent team dynamics.

In a new project where the team might start small and grow over time, it's especially important to keep in mind the following two principles: "Business people and developers must work together daily throughout the project" and "The best architectures, requirements, and designs emerge from self-organizing teams" [Manifesto for Agile Software Development 2001].

In his book *The Five Dysfunctions of a Team: A Leadership Fable* [Lencioni 2002], Patrick Lencioni describes the problems most likely to arise. Looking at the role of the individual within a team, he defines the five dysfunctions as

- *Absence of trust*: Unwilling to be vulnerable within the group

- *Fear of conflict*: Seeking artificial harmony over constructive debate

- *Lack of commitment*: Feigning buy-in for group decisions, creating ambiguity throughout the organization

- *Avoidance of accountability*: Neglecting to challenge peers on counterproductive behavior that sets low standards

- *Inattention to results*: Focusing on personal success, status, and ego before team success

Individuals often feel defensive at work due to an underlying feeling of vulnerability. Leaders who are not afraid to reveal their own vulnerability can encourage others to do the same. Author Lisa Bodell shows how teams can benefit from the increased openness that results. Vulnerability is not a sign of weakness, she states, but "a huge source of strength for a leader" and is, according to experts, a powerful tool for building genuine trust "by providing a safe space in which to discuss, experiment, fail, and innovate" [Bodell 2017].

Rockstar Programmer vs. Rockstar Team

As Scott Hanselman points out in "The Myth of the Rockstar Programmer," treating individuals as rockstars does nothing to enhance teamwork. He explains that giving a programmer this title actually harms the team for the following reasons:

- It creates unreasonable expectations for the others.

- It belittles and demotivates the rest of the team.

- The "rockstar" may somehow believe it's true.

Any team of professionals will match a normal distribution curve to a greater or lesser degree, and software developers are no different. The majority will be the average but good senior developers, there will be some junior team members with potential, and there'll be a few fantastic outliers (not to mention any others that are hopeless!).

People sometimes say that one rockstar can do the work of ten average engineers. That's sheer nonsense. It's a bit like saying that one very strong and skilled rower can row a boat designed for a crew of eight oarsmen. What really makes for a great team is a diversity of perspectives and experience. A *rockstar team* is definitely the thing to go for. Bring together some thoughtful and experienced architects with a few enthusiastic and energetic junior engineers, and you'll have a good mix. If you still feel you must call someone a rockstar, then make it the team leader [Hanselman 2013].

At the same time, in most teams you'll probably find there's at least one team member who really stands out. They may have a deep knowledge of the business domain or have the technical expertise or even both. Ideally, they enjoy actively sharing their knowledge and encourage others to deepen theirs too. Such team members are incredibly valuable. Nonetheless, a team of motivated individuals will generally outperform an expert operating in isolation.

Stakeholders

Another important task early in the project is to identify all the stakeholders. In simple terms, a stakeholder is someone who will be affected by the software project. Some of the stakeholders will be core members of the project team. Other stakeholders may impact the outcome of the project without being a user or project member.

The number of stakeholders required for your project will depend on the size and budget as well as on the departments and systems impacted. The typical stakeholders will include the sponsor, the project and/ or program leader, senior and line managers, project team members, users and/or user representatives, product managers, and subject matter experts. Further, there are business and technical support teams responsible for operations once the system is live. Other specialists include software enterprise architects, risk officers, security engineers, and auditors. Ideally, you'll be able to assemble the right people early in your project and brief everyone on the upcoming work. That way, you'll allow them to arrange the time and resources needed, even if the full detailed planning won't be available until a little later.

The Right People for the Job

So far in my career, I've had the luck to work with many smart and incredibly talented people, both on the business and technical sides. For me, the best ones are not those who are just smart, but those who also behave generously toward others. They share their knowledge, listen well, and appreciate that there are other clever people around. From a technical viewpoint, I admire people who go to great lengths to achieve the simplest solution possible, fulfilling the needs of all stakeholders at the same time. The skill of minimizing complexity throughout the software—as well as reducing dependencies and ensuring the system has low maintenance needs once it is productive–is a true craft, I feel.

Recruiting new team members is a challenging task. Unfortunately, not all recruiters and managers know how to select the best kind of person for a particular role.

A frequent mistake is to list an abundance of qualifications and must-haves in the job specifications, just to be on the safe side.

Such high requirements scare off some of the very people they're looking for. Especially for IT positions, when the recruitment agents have no technical background, they can't distinguish between the technical skills that can be learned in a week and other skills that take months to acquire. Too often have I discussed open internal positions with colleagues, and the description in the job posting rarely matched our understanding of the role in terms of which skills were required. Highly talented people are less likely to apply if they meet only 80 percent of the criteria, and this is due to the *Dunning-Kruger effect*. This term refers to a character trait sometimes found in persons of high ability who tend to underestimate their relative competence. Therefore, only the most important requirements and those that would take the longest time to acquire should be listed as mandatory. Often, the ability and motivation to learn and fit in with the team, together with the right soft skills, is more important than a few missing technical skills. Incidentally, a few fantastic developers I've worked with over the years were self-taught programmers. In most companies, however, they wouldn't even have made it to the first interview.

Another danger of listing too many required skills is that an over-qualified person may end up getting the job. They will quickly feel dissatisfied and look for a more challenging role elsewhere. In one instance, I saw how the members of a business team came and went in rapid succession—this seemed odd because the boss was a very likable and competent guy. The team was under a lot of pressure to improve the data quality, and although the job description became longer and longer, none of the people hired proved effective. Because the work was quite repetitive, a detail-oriented individual with good concentration skills was required. The successful candidates were overqualified. They soon became bored and lost interest in the job, frequently missing the data errors. Not all problems can be solved by adding further qualifications to the list.

Ideally, you need to find developers for your team who not only have the right technical skills but are also willing to learn about the specific business domain. Eric Evans states in his classic book *Domain-Driven Design*, "Most talented developers do not have much interest in learning about the specific domain in which they are working [...]. Technical people enjoy quantifiable problems that exercise their technical skills." This is a shame because the challenge of modeling the specific design is also a very interesting part of the job and as Eric writes, "The messiness of most software domains is actually an interesting technical challenge. [...]

There are design techniques that can bring order to a sprawling software application. Cultivation of these skills makes a developer much more valuable, even in an initially unfamiliar domain" [Evans 2004].

Team Pitfalls

Among the hundreds of people I've worked with, I've enjoyed collaborating with the large majority of them, and it's often been a lot of fun. I've noticed though that certain habits or traits in the minority can make teamwork difficult. If individual shortcomings have a considerable negative impact on the work quality and the team spirit, then I firmly believe they should be addressed directly and in private. Here are some of the behaviors I've witnessed that adversely affect teams:

- *Working selectively*: People who do only the work the bosses ask them for, making a good impression on all the bosses but delaying everyone else's tasks. This habit can be hard for managers to spot.

- *Overcritical of other people*: These are the ones who never say anything positive and can really drag down the project morale.

- *Racing through tasks, frequently switching tasks, and multitasking*: People don't realize they're generating more work both for themselves and for others by working in this manner. Discussions and clarifications will be required to correct their mistakes. Research shows that those who try to do too many things at once or to resolve problems too quickly are frying their attention span.

- *Lack of empathy*: An inability to see things through the eyes of others means these types only see the solution from their own specialist perspective. They cannot, for instance, understand why an end user finds their solution difficult to understand.

- *Lack of broad interest*: Certain IT people are entirely absorbed in their own technical challenges and pay no attention to the domain, the subject matter, or the company business. Likewise, some business types collaborating in IT projects refuse to take any interest in the IT side.

In my experience, the last two characteristics especially have resulted in bad decisions, delays and additional loops, escalations, and extra meetings. People who behave like this also tend to throw everyone in the same bag, as mentioned earlier, making sweeping statements like "The business people don't know what they want" or "The IT team is too slow and too expensive as always."

An awareness of these types of behavior will help you identify and address such problems in order to get the best out of your team.

Next, we'll look at the task of defining the requirements of the software, based on the business goals of the project. These decisions, which involve close collaboration between all the teams, will dictate the functions of the software as well as its appearance.

Requirements

Simply put, a requirement in software development represents the need of a stakeholder or user. This need usually adds some form of value to the system or solves a problem therein. Beyond the user or stakeholder, a contract could also necessitate a requirement, as could compliance with a law, meeting a standard or norm, etc. If the business departments and stakeholders are not in agreement, however, the task cannot be implemented successfully. It is therefore crucial to align the different needs, ideas, and opinions of everyone involved in the project, ideally before any code is written.

The documentation will describe these requirements using natural language written in prose. You'll find that diagrams, charts, and other forms of visualization help bring greater clarity to the descriptions. These visual elements can be very useful in illustrating the needs more distinctly and precisely, especially as prose can sometimes be vague or ambiguous.

> *English, or any other human language, is not naturally a precision instrument for such definitions. Therefore, the manual writer must strain himself and his language to achieve the precision needed.*
>
> —[Brooks 1995 p. 63]

The projects I've worked on where the requirements were clear and solid moved along very smoothly and saved a lot of time.

Defining the Requirements

The requirements document is divided into *functional* and *non-functional* requirements (NFR). Functional requirements mainly describe the behavioral features of the software being built. The visible parts of the software usually come into this category, including the business rules. Non-functional requirements focus on the technical system operations such as performance, availability, security, etc. A non-functional requirement could be the need to load a web page in 2 seconds or less, for example.

The business is tasked with defining the functional requirements by formulating the business goals to be achieved. Copying missing address data from System A to System B, for instance, is not a requirement, but may be one of many possible solutions. This shift from thinking in terms of solutions to describing the actual issue is critical. You should never formulate requirements as solutions—it means that important considerations and alternatives may be overlooked. In other words, the requirements should be technology and system agnostic—this means expressing the business needs in a way that's independent of existing systems.

A recent documentary showed the planning work for a new link between Germany and Denmark across the Baltic Sea. The lead engineer, Sandra Akmansoy, explained how the government approached her firm. "They said they wanted a bridge, but we asked: 'Do you want a bridge or do you want to cross the waterway?'" As she says, it's all about questioning things [Seewald et al. 2018]. A tunnel proved to be more suitable and cost-effective for this project's requirements. This example clearly shows how the underlying goals need to be identified first.

In larger projects, you'll need one or more dedicated requirements engineers to identify the needs of the new system. Whether the task is large or small, identifying all the requirements is of paramount importance for the project outcome. The items specified in the requirements document dictate what shall and shall not be built. The person(s) gathering and writing the requirements must therefore have a solid understanding of both the business itself and the processes within the project scope. They also need to translate these needs into quantifiable, prioritized requirements that express the overall intent of the project.

The goal here is to create a detailed model of the business domain.

Definition of awesome is a term often used in software projects. The phrase represents a challenge for team members to imagine the best solution possible for a given problem or feature, regardless of whether the goal is achievable or not. I strongly recommend everyone, including the business people, to give their definition(s) of awesome in a project. The goal might just be possible. Why not place your definition of awesome on the board for everyone to see? It could prove to be an inspiration! Brainstorming and mind mapping are also widely used methods that are helpful in identifying innovative features. A further interesting concept that a colleague of mine swears by is *brainwriting*. This is like brainstorming, except that individuals write the ideas down. He feels this method works better for the more reserved types.

The team should use various approaches to gain an all-round view of what the system should do. Possible approaches include interviews, surveys, floorwalking, and sitting with users while they work. You should prepare the information in annotated visual form as far as possible. Research into neural science shows that people understand information represented visually more quickly and memorize it more effectively than information documented in natural language (e.g., [Glass & Holyoak 1986] and [Kosslyn 1988]).

Although somewhat harshly expressed, the phrase "garbage in, garbage out" is very true of the requirements document. Providing incorrect, incomplete, or low-quality requirements will only produce faulty output.

Once the underlying needs have been defined, the teams can turn their attention to the existing systems. Note that there's an important distinction between the *system*, the *system context*, and everything outside the system context (Figure 13-1).

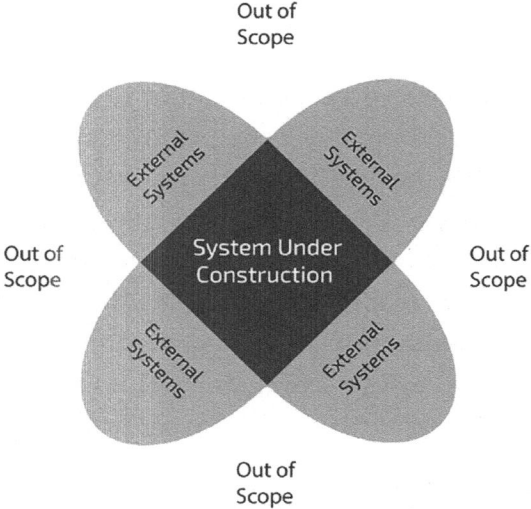

Figure 13-1. *System scope*

The system in this case refers to the software application the project is constructing. The system context, on the other hand, defines the surrounding dependencies that will influence how the software is built. These dependencies are labeled in the preceding diagram as "External Systems." Anything that lies beyond the system context is defined as out of scope and irrelevant for the project and should therefore be documented as such.

The analysis of an existing system is a rich source of information—the current configuration usually reflects the accumulation of many business decisions. You need to ask: are any of the features no longer used, or can any of them be improved in the new system? As part of the analysis of the existing system, be sure to look carefully at the current production issues. The team supporting the application may have had years of hands-on experience and could therefore offer a long list of useful improvements. The existing system may also deliver reliable statistics that can help you make qualified decisions on new features and changes. How often is a feature or page being used? Combine the statistics with valid reasoning.

Why was a feature not used often? It could be badly documented, too slow in loading, or only needed periodically, such as a monthly report. On the other hand, of course, the feature could just be too complicated or not very useful.

In one project, we were discussing at length how best to add functionality to a feature—it was going to be a rather complex task. As we weren't moving forward or finding a simpler solution, one of the developers suggested we check how often the feature was actually being used. Once the business department realized that the feature was rarely utilized despite being easy to use, they changed their mind and asked us to focus on other features instead.

Identify any data contained in the current system that needs to be transferred to the new environment, and ensure this is planned appropriately. This transfer of data is called *data migration*. During this process, the team may need to adapt the structure of the data to a new format.

The authors of the book *The Pragmatic Programmer* suggest sitting with an end user for a week. "Management will give you one view of how things operate, but when you get down to the floor, you'll find a very different reality—one that will take time to assimilate" [Hunt & Thomas 1999].

As the quality and format of the requirements will have a significant impact on all the successive work steps, certain criteria should be met. To ensure a high standard, take care that each requirement

- Addresses only one item
- Is complete
- Is consistent
- Is atomic (broken down into the smallest details possible)

- Is clearly formulated

- Is verifiable

- Has acceptance criteria

Many books and tutorials refer to requirements gathering as an early phase of the project. [...] Gathering implies that the requirements are already there—you need merely find them, place them in your basket, and be merrily on your way. It doesn't quite work that way. Requirements rarely lie on the surface. Normally, they're buried deep beneath layers of assumptions, misconceptions, and politics.

—[Hunt & Thomas 1999]

A useful way for checking the quality of the objectives, and hence the business requirements, is by using the acronym SMART to ask if a requirement is

Letter

S

M

A

R

T

Meaning

Specific

Measurable

Achievable

Relevant

Time-bound

Do all your requirements fulfill these five criteria?

Another popular term in the software development world is YAGNI: "you aren't gonna need it." This means you should build only the things that are required immediately.

Finally, the project team must ensure that all the stakeholders have read and fully understood the requirements document before they give the sign-off. The team also needs to resolve any conflicting or contradictory requirements before beginning development. The acceptance criteria at this stage will later form the basis of the user acceptance test. All the requirements should be consolidated and stored in one central place when handed over to the development team. They shouldn't have to hunt for facts and decisions that are buried in long email chains or meeting minutes, for example. Also, the team should be allowed to send back any requirements that don't meet the necessary standards.

Waterfall vs. Agile Requirements

You've already been introduced to the Waterfall and Agile approaches in the "Methodology" section of Chapter 12. To follow on from this, we can now look at how the chosen methodology will influence gathering the requirements. The requirements analysis is strongly affected by the methodology, although the main overall goals should be clear for both ways of working. When using an Agile approach, the requirements are addressed in smaller parts with the goal of reducing complexity and avoiding *analysis paralysis*—which means that a project is stalled at the analysis stage. The features are discussed and prioritized based on the business value of each one. Depending on its complexity, a feature can be broken down into even smaller parts. One further task in the Agile methodology is to track the requirements to avoid requirements *creep*, also called *scope creep*. This term describes an uncontrolled growth in project scope.

It isn't always practical to specify 100% of the requirements or designs upfront, but most projects find value in identifying at least the most critical requirements and architectural elements early. Iterative approaches are usually a better option for many reasons, but an iterative approach that ignores prerequisites can end up costing significantly more than a sequential project that pays close attention to prerequisites.

–[McConnell 2004]

Unlike Agile, the Waterfall approach expects most or all requirements to be collected upfront.

Requirements are the things that you should discover before starting to build your product. Discovering the requirements during construction, or worse, when your client starts using your product, is so expensive and so inefficient, that we will assume that no right-thinking person would do it, and not mention it again.

—[Robertson & Robertson 2013]

Also, regarding the Waterfall methodology, *Code Complete* recommends that you "estimate the time for the rest of the project after you've finished the requirements" [McConnell 2004].

User Requirements

Holding user requirements workshops can be very helpful in gathering these requirements. Lighten things up and add some fun to these workshops by using a large buzzer-type button, for example (Figure 13-2). The stakeholders take it in turns to say, "When I press this button, I want to export my monthly sales figures to a PDF file" or "When I press this button, I want the system to calculate the correct interest." The buzzer is symbolic, and not everything in the application will be triggered by a button.

However, by activating both sides of the brain and taking turns, group members can be actively involved in contributing a large part of the user requirements. If it doesn't feel like work, then that's exactly what you want.

Figure 13-2. *Break the ice and add a little fun with a gimmick*

One activity that will engage everyone, and is very useful at this stage, is to experiment with roleplay to represent the different types of typical users. This can be great fun too! Project members adopt personas, and ascribe to everyone a name, age, profession, sex, and any other characteristics as appropriate. This method is used to create a scenario for each persona to gain insights into how they might interact with the system. The objectives, likes, and dislikes of each persona can be documented and visually displayed on the project room wall to stimulate ideas.

Use whiteboards, flipcharts, cards, and other aids to illustrate the main points so that they're clear to everyone.

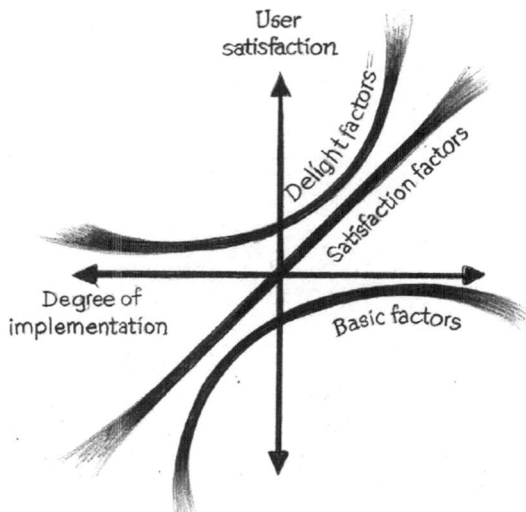

Figure 13-3. *The Kano model categorizes user expectancies*

According to the Kano model (Figure 13-3), user expectancies involve three different factors [Kano et al. 1984]:

- *Basic factors*: The functionalities the user feels go without saying

- *Satisfaction factors*: The functionalities the user explicitly requests

- *Delight factors*: The innovative functionalities that pleasantly surprise the user

The user requirements also need to be categorized and prioritized according to the order of importance. As such, these requirements will fall into one of three categories: legally binding, strongly recommended, or nice-to-have. Some business representatives struggle with these business decisions. Scorecards can be used as an aid to categorizing appropriately in terms of complexity for IT and importance for the business side.

I've noticed that most users and stakeholders tend to think in terms of the standard business case, but the system also needs to allow for rare cases. In one example, the user said, "We just need to store one address for each customer." I asked, "Do you never store more than one address per customer?" The user then added, "Oh, actually, we do have a few exceptions, and for one client we store six addresses, but that's hardly ever the case." This kind of input is very valuable, as the final solution may need a design for displaying a list of addresses instead of just one. These addresses may also need headings such as "home address" or "business address." Software projects are full of exceptions, and these need to be identified.

The handling of untypical user behaviors and failed cases should be discussed too. What should happen if an online payment is not successful? What if an employee accidentally sends a message to all the customers—can the system prevent that? There will nearly always be some situations where the usage will vary from the standard cases; hence, some thought should be given as to which ones will be supported. Good requirements documents provide for the unlikely and unexpected user behaviors as well as the likely ones.

One kind of feature request I strongly discourage goes something like this: "Just make the new xyz feature identical to the existing abc feature—they're practically the same." When the requestor tries to save time by failing to specify any details, this kind of "shortcut" can lead to a lot of issues and additional reworking. I've fallen into that trap myself. Although the new xyz feature may be very similar to the abc one, there will be some critical differences that the requestor didn't consider. The existing feature could be used as a template for the new feature, but the content should be reconfirmed on every point. "Repeating all the essentials tends to make manuals dull reading, but precision is more important than liveliness" [Brooks 1995 p. 62].

Something often overlooked initially is the order in which data is listed on the user interface. A simple alphabetical ordering of elements won't always cut it. The business may, in fact, require a customized order that in turn will require manual maintenance if a new item is added. Who will be responsible for managing this sort order? The maintenance of such an ordered list may require a separate pop-up window. Hence, one small detail can easily entail more work than expected.

User Stories and Use Cases

User stories and *use cases* are two common methods of evaluating the user's needs, and these form part of the requirements. The former is based on a simple, informal sentence, whereas the latter is more formal, usually covering a much broader scope than just the user needs. The choice between the two will depend on the project, the collaboration, the level of formality, and whether any upfront research is required. Accordingly, user stories are generally better suited to informal business projects, whereas use cases are mainly used for formal projects with stringent documentation requirements. In some projects the two methods are complementary and are used in tandem.

User Stories

The user story is usually evaluated in an interview or in a workshop. It involves the idea that a *person* wants to be able to perform a specific *action* so that *a goal can be achieved*. The user story is therefore based on a simple sentence such as this: "As a librarian, I want to be able to scan a book so that the customer and I know when the book has to be returned." During the evaluation sessions, user stories are noted down on cards, giving a rough idea of the goals, but without going into detail at this point of how these goals will be achieved (Figure 13-4).

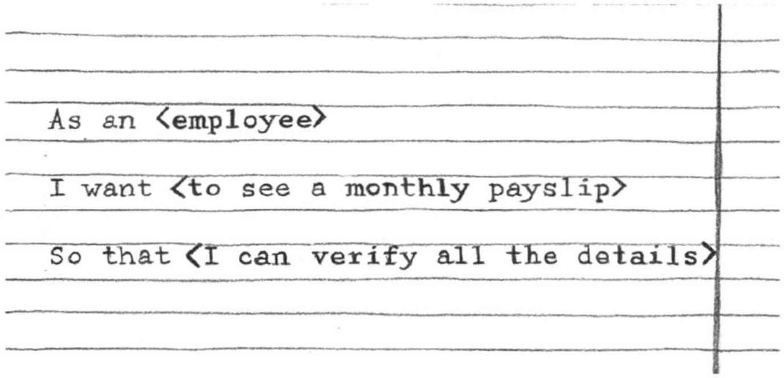

```
As an <employee>

I want <to see a monthly payslip>

So that <I can verify all the details>
```

Figure 13-4. *An example of a user story card*

Collecting user stories helps to provide an overview of what a system should do. These requests can then be assessed according to business value. The next step is to analyze in more depth the features to be done first. The task of defining the acceptance criteria is often done during this stage—the acceptance criteria describe the point at which a new feature will meet the requestor's expectations.

Visual storytelling is another interesting approach that can also be used to understand the users (Figure 13-5). These visualizations go beyond a single viewpoint to show the relation between the individual steps from a user's perspective and their potential behaviors [Bradd 2016].

Figure 13-5. *Illustrating the user's behavior using visual storytelling can help identify the requirements*

Use Cases

Use cases document the flows of system usage, describing the processes a user goes through step by step to complete one or more goals. These flows are expressed in written form and are enhanced by diagrams and illustrations of models that are easy to understand. The use cases describe the features of a system from a "user" perspective, where such a user could be a person, a role/job function, or even another system. The use cases also describe the things that can go wrong along the way. A typical use case includes the elements given in the following.

Elements of a Use Case

Depending on how formal the use case needs to be, some of the following elements can be used. In very formal projects, even more fields will be required:

- *Name*: A clear naming of the goal of the use case

- *Brief description*: A short paragraph of text describing the scope of the use case

- *Actors*: The types of users who perform a behavior using the system

- *Stakeholder*: Someone with vested interests in the behavior of the system

- *Primary actor*: A stakeholder who initiates an interaction with the system to achieve a goal

- *Preconditions*: What must be true or must happen before the use case can run

- *Triggers*: The events that cause the use case to be initiated

- *Main success scenarios (basic flow)*: Use cases in which nothing goes wrong

- *Alternative paths (alternative flow)*: Variations on the main theme that result from things going wrong at the system level

A UML use case diagram depicts various use cases in the simplest form and shows the activities the different users expect to achieve (Figure 13-6). UML, which stands for *unified modeling language*, is a general-purpose language used to visualize system concepts.

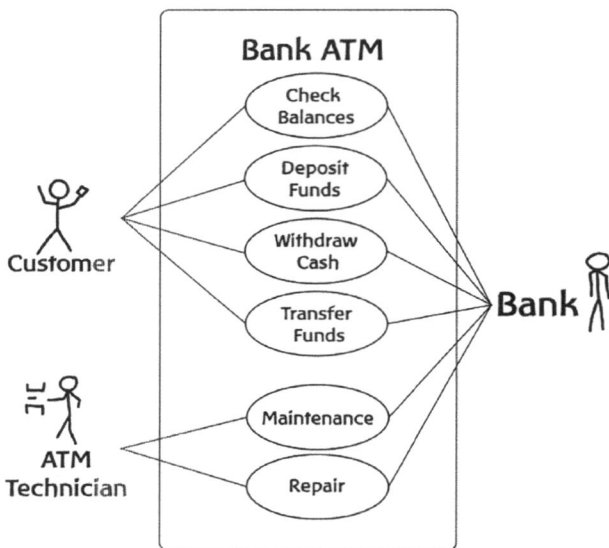

Figure 13-6. *A typical use case diagram*

Decide and Be Precise

Implementing software that's based on laws and regulations can be very difficult if there's any vagueness. I know this from experience. The legal teams need to be vigilant in their decision-making, for instance; otherwise, imprecise information will be passed on to the development team, who will probably just return the requirements along with lots of questions. Alternatively, the developers may work using certain assumptions, only to discover later that the solution is incorrect.

Whatever the cause of vagueness in business cases, I strongly recommend that the business teams and subject matter experts evaluate and interpret these cases. They should then specify the requirements as concretely as possible. Don't expect to have full details of every requirement upfront, but be prepared to make some tough decisions.

As outlined earlier, companies should build a culture of responsible decision-making. Simply passing the buck causes big delays in any undertaking. An inability to make well-balanced decisions is often rooted in the fear of getting things wrong and possibly being shamed. Of course, important decisions should not be made hastily or without involving all the teams affected. Due to the preciseness of software, every detail must be specified—any vagueness will only lead to problems sooner or later.

Your Dependency on Others

Every project will have some form of technical and human dependencies that can pose a substantial risk to the project. These need to be managed well to avoid delays. Such dependencies can have a major impact on testing too. Your project may even become date driven because of such external commitments.

With new systems, the dependency is often one-sided—your project may be dependent on other systems and teams, but they are not dependent on yours. Likewise, the infrastructure teams can do their job without you, but not vice versa. Any cases where software development teams and IT infrastructure are not yet working closely together need attention. As already mentioned, one effective way for getting these IT teams to work together better is known as *DevOps*—where development and operations teams work together to achieve more business value. The management in both these IT departments needs to pay close attention to the dependencies—otherwise, these can drastically slow down the project and create budget overruns.

Are there any systems that will affect your timeline or that cannot make any changes due to other important projects?

Further, some dependencies may not be clear at first but will show up later in the project. These are more likely to appear if new requirements come up while the project is underway. A UML activity diagram depicting the structure, behavior, and interaction of the system can assist in identifying these dependencies. These diagrams provide a visual representation that will highlight the dependencies both within and outside the system. Figure 13-7 is an example.

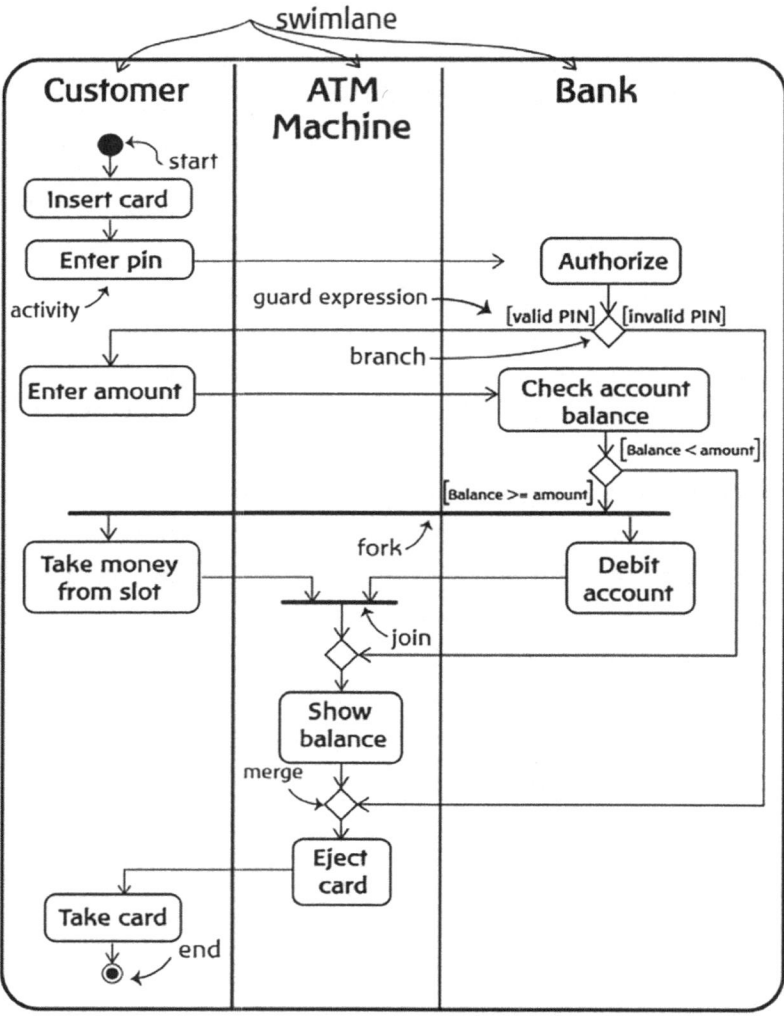

Figure 13-7. *A UML activity diagram showing the simplified process of a customer using an ATM*

Technical dependencies can range from technical infrastructure like the servers, communication systems, databases, and so on to systems delivering or receiving business-relevant data. Dependencies on persons, meanwhile, can, for example, be due to critical stakeholders having had no time to define their requirements.

One type of dependency that can add complexity is called a *moving target*. This term refers to a situation that keeps changing when connecting to another system or receiving its data. A test might be successful one day, yet a few days later, the same test might fail because something in another system has been changed. When this scenario keeps repeating, it becomes very hard to identify the real issues in either system. Ideally, there should be as few moving parts as possible.

Your Impact on Others

The *impact analysis* is a dedicated step that aims to identify all the systems and processes affected by your project. Before creating a new system or making changes, you'll need to put a lot of thought and planning into recognizing the potentially widespread consequences.

Impact analysis is the technique of identifying the possible repercussions that can arise when making a change or adding something new. The data sent by your system to the surrounding ones will usually be the starting point. The fact that these systems may send your data further down the line is something that is frequently forgotten. It's always vital to contact the whole chain of affected systems. Once you have a complete list of the systems and teams affected, you'll need to discuss the actual positive and negative impacts of the change with these teams.

Keep on top of the situation by proactively informing other teams and collaborating in advance. Companies should actively encourage and enable this proactive approach. If too many obstacles are placed in the way of planned work, it may discourage projects from performing full impact analysis. This lack of coordination will lead to emergency changes, reduced testing, and more bugs and can mean some other important work is delayed.

System Interfaces as Contracts

The word *interface* is used in many different areas of software. The meaning I'm referring to here is basically the connection between two systems. An interface can also be understood as a contract, according to which certain data and data structure will be requested and returned.

System A may need customer data from System B, for example. To retrieve this data, System A could send a unique identifier (e.g., ID=11244), and System B would respond with all the predefined fields (ID=11244, Firstname=Sara, Familyname=Cornell, DateOfBirth=02051980, Email=, etc.).

These contracts between systems, usually called *interface agreements*, may also contain some useful details on how the systems will act if the data isn't found and on how errors are handled. Additionally, these agreements will also include details of the required performance and availability.

Next, we will consider the quality of the software product. This is another factor you'll need to decide on that will largely depend on the time and resources available.

Quality

The idea of quality tends to be subjective. If questioned on the subject, everyone will say, yes, they want better quality. But the matter of quality must always be weighed against time and cost. The two main indicators of software quality are, firstly, adherence to the functionality expected of a piece of software and, secondly, whether the non-functional requirements are met.

To briefly recap, a functional requirement could mean adding a product to the shopping cart in an online shop. Can this customer goal be met? Do the cart buttons work? Does the quantity × price – voucher = the correct total? And so forth. One of the non-functional requirements is website performance. How long does a customer have to wait between clicks for the website to respond? Both aspects of quality are important to keep the customer on the website and to complete the transaction.

> *If you emphasize quality at the end of a project, you emphasize on system testing. However, testing cannot fix design-time mistakes. If you emphasize quality in the middle of the project, you emphasize on construction practices. If you emphasize quality at the beginning of the project, you plan for, require and design a high-quality product. If you start the process with designs for a Pontiac Aztek, then you can test all you want, it will never turn into a Rolls-Royce.*
>
> —[McConnell 2004]

In most cases, just making a checklist longer doesn't improve the quality. Instead, the indicators of quality should be well selected, measurable, and easy to understand.

Data Quality

Data is considered high quality when it accurately represents the real-world construct. The data should be appropriate for the specific use and within the given context. Throughout my career, I've known this topic to be one of the issues that end users have reported most often—it is one of the top support issues. The impact of poor data on businesses incurs huge costs. According to some studies, these costs reach more than US $3 trillion annually in the United States alone. Letters and parcels being sent to the wrong addresses are one of the many outcomes.

Despite causing disruption to the business as well as the IT support, data quality issues are often not sufficiently prioritized. Adding tailored data guards across all applications during input can help reduce this problem considerably. Although not very costly, this validation of data is sometimes implemented inadequately—mainly because companies barely recognize the wider implications of the issue.

In a recent example of bad data flow, a German bank accidentally transferred $35 billion to a stock exchange instead of a much smaller amount. Such mistakes can be mitigated by using more sophisticated validation, which I'm sure the bank will now implement, following this embarrassing and potentially costly incident [Canny 2018].

It's vital to first ensure the data is input correctly at the source—accuracy can be achieved by using technical restrictions when the data is typed. An obvious example is checking that an email address has the correct format. User training and monitoring the quality of the input data are further important data tasks. Additionally, any external data delivered can be systematically checked upon receipt.

As briefly mentioned earlier, at one company I experienced a central system receiving poor data from an external data provider. Even though this central system supplied many surrounding systems with data, no data guards were in place. Surprisingly, the team supporting the central system did not feel at all responsible for the low quality of the data—it was delivered from an external source, and they were just passing it on. Hence, every system that received this data had to implement its own data guard logic to manage the issue.

Data profiling and *data cleansing* are both methods for improving inconsistent data. The first involves gathering statistics about the existing data and then examining it. The data can subsequently be analyzed to assess whether the quality meets the standards expected. Data cleansing is then done to fix any incomplete and/or incorrect data.

IT Audits

Most companies are subject to external audits, and some may undertake internal audits too. Typically, audits focus on overall quality and process adherence. More specific audits cover the IT infrastructure, data, and software. The teams responsible are obliged to deliver any information demanded by the auditors. They may require detailed information on documentation, access management, security, segregation of duties, data integrity, or any other area they see as relevant.

As you may have noticed, all the different parts of software development are interrelated. Security is another factor that will affect the software, including the functionality and the data, confidential or otherwise. The next section will give an overview of how you can protect your software and the data it contains, sends, and receives.

Security and Data Access
Security

IT security remains a hot topic, especially in the wake of global attacks, the scale of threats, and the media attention given to this matter. A serious attack on a company can, after all, result in direct financial losses, stolen data, and reputational damage. In many countries, if customer data has been compromised, the company concerned is obliged by law to inform its customers individually about their stolen data.

Security involves a range of considerations in every IT project. Assessing and dealing with this wide area is not an easy task. As risks can come from internal or external sources, software creators need to anticipate every possible way that someone could break in, as well as guarding against any new threats that could evolve over time. Various organizations regularly provide lists showing the current top 10 and top 25 global threats and so on.

By utilizing automated tools to check the code, you can ensure these threats cannot be exploited in your code. Also, specialized companies can test your software or website for vulnerabilities.

A dedicated IT security team will most likely be available in mid- to large-sized companies to support your project. The increased sophistication of the threats means the project needs to place every part of the software and the infrastructure under scrutiny. The essential thing at this stage is for your company's security team to review the project and help deliver a *security concept* document.

Data Access

Defining which people should see which data is a crucial matter too. It requires a closer involvement from the business departments than a more technical area like security. A common approach to managing the access to different areas of an application is to create specific roles—this is particularly useful for applications with many different usages and access levels. A user's account is then assigned to one or more of these roles. It's generally much simpler to manage these groups than attempting to manage hundreds of users individually.

A basic role concept could include an admin role for the support team as well as read-only and read-write roles for end user access. A public website might differentiate between public areas, member roles, and admin roles. Larger systems typically have a more sophisticated role concept.

Data access goes beyond the data that is visible on the screen. Any reports generated, or data exported to other systems, need to be carefully analyzed too. A friend of mine, who works for a hospital, told me of the mess-up they had had with data there. The patient data system was very restrictive in allowing employees access to patients' personal data and health history. Designed to show data on a need-to-know basis, the system worked very well—until one employee discovered that the billing system wasn't as restrictive as the patient data system. It turned out that every employee with access to the billing system could see the health history of all the patients, including that of all employees who had previously been patients in the hospital themselves.

The task of software design, which is conducted by the software engineers, should involve taking such privacy factors into account to avoid this kind of scenario—the task concerns the structure of the various components and the interaction between systems. A brief overview of the software design task follows.

Software Design

The way software is designed naturally changes over time and as trends come and go. In the early days of computing, *monolith* systems, where a single system could be responsible for a wide range of tasks, were common. A more common approach today is to use microservices, where the functionality of each service is fine-grained.

High-Level Design

The high-level design (HLD) or high-level solution design document is the blueprint of translated requirements for the actual development of the software. This document gives an overview of the system, specifying the main components to be developed and any external systems that will be connected. A solution architect or senior developer may write the HLD, following consultation with an enterprise architect. The solution architect is usually responsible for the design of specific software applications. The enterprise architect, on the other hand, is responsible for the overall company-wide standards and interaction between systems.

The HLD should offer a complete portrayal of the whole system but with minimal technical jargon—to allow reviews by all the stakeholders. For the purposes of clarity, the HLD, like the requirements document, will make use of diagrams. These include a detailed architecture diagram alongside a text written in prose. Additionally, you should find the security, application, database, and technology architecture outlined in this document.

> *A high-level design provides an overview of a solution, platform, system, product, service or process. The idea is to briefly mention every work area involved, clearly showing how the ownership of the more detailed design activity will be delegated, whilst also encouraging effective collaboration between the various project teams.*

> [High-Level Design–Wikipedia]

Low-Level Design

The low-level design (LLD) documents to be produced later will be based on the HLD and will describe the solution to be built at a far more granular and technical level. Separate documents are usually created for the different requirements or features to be worked on. The processes of estimating and prioritizing may be done prior to producing the LLDs—this ensures that the features will be included in the upcoming version of the software.

These documents focus on the internal logic of the software code and on how the details will be implemented. This step is often the concern of the software development engineers and their team leader. The documents are typically then validated by the solution architect and adapted until both parties are satisfied. In some cases, the enterprise architect will also need consulting.

These design documents need to strike a balance between sufficient granularity and over-engineering. There isn't much sense in filling this document with the actual code, which will be written later anyway.

One common approach is to use a *pseudocode*—this uses the structure of a typical programming language but is written in natural language. The pseudocode makes the document easier to read and discuss than conventional code. It's a sort of in-between language that is used before the writing of the actual code. A very simple example of pseudocode would be

If learner driver has made less than 5 mistakes during the driving test
print "passed"
otherwise
print "failed"

Although you'll already have done some initial planning at the beginning of the project, more detailed planning is required before the actual development process begins. Next, we'll go over the areas you need to focus on now.

Detailed Planning
The Trouble with Estimates

Even for people with many years of experience in software projects, estimating timescales and costs is hard to get right. Estimates are, after all, only estimates and should be treated as such. Often a team's time estimates are misinterpreted as hard dates. If these are not met, and the team is subsequently criticized for having been overoptimistic, then they may define their future estimates too pessimistically as a result. Ideally, estimates should always be as realistic as possible, taking all uncertainties into account and allowing leeway for any unexpected factors. To encourage realistic estimates, any delays should be discussed and handled objectively.

All estimates are wrong.

The task of estimating times for implementing the various requirements is a team sport. Understanding the technical complexity of a planned feature is one of the strengths of an experienced development team. Each team member will have different knowledge and experience—this helps greatly in discussing how long a specific feature will take to code. If estimates among members vary greatly at a team estimation session, they'll need to discuss the reasons for these differences and come to an agreement. The team should also contemplate the potential for experimental features. Further, an estimate based on concrete requirements with little chance of change, in contrast to vague requirements with a high chance of change, should be labeled accordingly. This clear labeling will give an overall indication of the uncertainty involved in the upcoming software release. The estimates should also allow for any dependencies on other teams and other systems—these external factors should have been identified as part of the requirements.

Is the other system developed by an external company? How fast can they deliver? What is their software release cycle?

Your business teams should be involved in the cost estimates too. Where the estimated cost of a feature is very high, together you may think of a more cost-effective way of achieving the same goal. In an initial feature request, you may rightly have requested an optimal or preferred solution but may now be prepared to opt for a simpler one rather than suffer a delay or cost overrun.

Prioritizing

Prioritizing, which is closely linked to estimating, needs to be a team effort too. At least one representative from each stakeholder group should attend these meetings. Whereas the estimates require the expertise of the development team, the prioritizing is based on business value. It is therefore largely a task for the business representatives. Yet don't forget that paying off technical debt should be considered as a prerequisite to adding more business value—it can prevent serious technical difficulties and delays a little further down the line. The teams could use a points system to prioritize requirements, according to both the business needs and technical complexity. By reviewing the main criteria defined for each requirement, they can indicate if it is a must-have, should-have, or a nice-to-have feature, thereby making the prioritizing much easier.

Prioritizing should not be influenced by which team or customer shouts the loudest for a new feature. Rather, these decisions should be based on the product manager's careful evaluation of the various needs. Some features may be useful for only a very small number of clients, and a fair distribution of effort to satisfy as many clients as possible will lead to better results in the long run. Prioritizing individual clients' requests instead of adopting a broader approach may simply lead the less vocal clients to take their business elsewhere.

The number of less intense clients may, in fact, far exceed the more insistent ones constantly asking for the specific new features required only by them. I've experienced product managers who routinely placed the loudest requests on top.

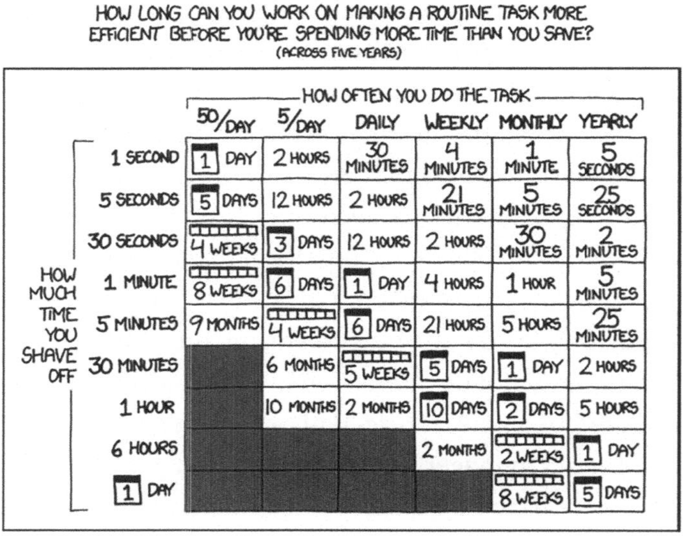

HOW LONG CAN YOU WORK ON MAKING A ROUTINE TASK MORE
EFFICIENT BEFORE YOU'RE SPENDING MORE TIME THAN YOU SAVE?
(ACROSS FIVE YEARS)

		HOW OFTEN YOU DO THE TASK				
		50/DAY	5/DAY	DAILY	WEEKLY	MONTHLY	YEARLY
HOW MUCH TIME YOU SHAVE OFF	1 SECOND	1 DAY	2 HOURS	30 MINUTES	4 MINUTES	1 MINUTE	5 SECONDS
	5 SECONDS	5 DAYS	12 HOURS	2 HOURS	21 MINUTES	5 MINUTES	25 SECONDS
	30 SECONDS	4 WEEKS	3 DAYS	12 HOURS	2 HOURS	30 MINUTES	2 MINUTES
	1 MINUTE	8 WEEKS	6 DAYS	1 DAY	4 HOURS	1 HOUR	5 MINUTES
	5 MINUTES	9 MONTHS	4 WEEKS	6 DAYS	21 HOURS	5 HOURS	25 MINUTES
	30 MINUTES		6 MONTHS	5 WEEKS	5 DAYS	1 DAY	2 HOURS
	1 HOUR		10 MONTHS	2 MONTHS	10 DAYS	2 DAYS	5 HOURS
	6 HOURS				2 MONTHS	2 WEEKS	1 DAY
	1 DAY					8 WEEKS	5 DAYS

Figure 13-8. *Is the task worth automating?*

Drawing up a chart, such as the one shown in Figure 13-8, is useful for calculating the value of automating a routine task. The chart will indicate whether the work will be worthwhile, as automation in some cases can involve more work than performing the task manually. For each routine task in question, multiply the number of times the task is performed by the time that would be saved through automating it. The resulting figure can then be measured against the time needed to develop and test it.

Test Planning

Every piece of software requires testing. Chapter 15 goes further into the details of this topic. The testing is a team effort involving just about everyone: developers, business analysts, subject matter experts, project leaders, end users, and, finally, dedicated testers and test automation specialists.

The creation of test cases, the preparation of test data, the testing itself, and then the documentation of all the test evidence require a substantial amount of planning and time. The testing will also include any connected external systems, and the corresponding teams will need to provide appropriate resources for fixing the issues if any of those tests fail.

As part of each requirement, the acceptance criteria should have been defined. The user acceptance tests will draw on this information to ensure the goals have been met. Planning the time for the end user testing is important too. Any project delays can result in the end user testing phase falling in the holiday season and causing additional delays.

Maintenance Planning

As an important project stakeholder, the maintenance team of an application can usually provide invaluable information on user preferences, user difficulties, usage statistics, and any previous maintenance troubles.

Regardless of whether there was a previous system or not, the maintenance team needs to learn the details of the new system being built early in the project. Building up maintenance to support a completely new system can involve a considerable learning curve and does take time.

Getting the maintenance right is a key factor in keeping users happy right from the start. At the same time, creating a system with low maintenance costs will keep everyone happy. Delaying discussions on maintenance until halfway through the project or later can result in problems. You may discover important conceptual requirements that should have been included.

Great architects and development teams will therefore work closely with the maintenance team right from the word go. Throughout the project the maintenance team will focus especially on access rights, on tracking the important steps in the systems (logging), and error handling. Any element in a system could fail or be incorrectly configured. Having the right information helps the support team quickly identify the root cause of an issue. Any area in the system where something is unclear or untraceable is known as a *black box*. For maintenance purposes, it's best to avoid leaving behind such black boxes.

During the development phase, you may need to plan for some setup and configuration to enable monitoring and analytics. These features give a view of the system's health as well as the user's behavior when interacting with the software. As such, both are very relevant to the business as well as to the maintenance teams (see the section "Monitoring and Analytics" of Chapter 16 for more details on this topic).

Even with the very best planning, unforeseen factors are almost certain to crop up and necessitate changes. I'd like to add a few words here on this topic.

Handling Changes

The ability to take everything in your stride and handle change well is crucial in every project. A change could be due to many factors, including changed circumstances, new laws, a better understanding of the problem, or technical advances. You should also be prepared for the higher costs that changes almost inevitably bring. As we've seen, an Agile approach is usually better suited for projects where a lot of changes to the requirements are expected.

Further, be sure that all changes are streamlined centrally through a dedicated person or a change control board, depending upon the project size. This board will ensure that any changes to requirements during the project are documented and coordinated with all the relevant stakeholders. The business departments should therefore refrain from requesting changes from the developers directly.

For each request, the nature of the change and the impact for each target group should be evaluated. Massive changes arising during a project can even lead to the whole project failing, as the changes will need to be made both to the old system and to the one under construction. Therefore, senior management or even executive support may be required to evaluate whether to place other initiatives on hold.

The project team also needs to estimate how change requests affect other requirements as well as the test cases. Additionally, a cost-benefit analysis will be useful in assessing whether a change is worthwhile.

Next, I'd like to point out some of the mistakes people commonly make during the project preparation stage to help you avoid them in your project.

Pitfalls—Part 1

To conclude this chapter outlining the crucial decisions you'll need to make for your project, we look at some of the bad decisions commonly made during this phase that you'll want to avoid.

The organization of teams will need adapting if the project grows considerably. A team of six can probably handle a lot of tasks informally, whereas 20 people may need splitting into two or three teams. Also, the communication between them will require a more systematic approach. Although "realigning teams to make them more efficient" may sound like a great idea, in complex environments it can throw everyone off track to suddenly reorganize teams in the middle of a project. It can also create a lot of extra learning curves. To avoid such disruptions, try to make only small amendments to teams when really necessary during a project.

One of the difficulties in software projects always comes from the fact that *the devil lies in the detail.* Some very low-level technical restraints can actually have a strong impact on high-level decisions, for instance. In one project I worked on, the goal was to display two images side by side, allowing the user to vote on their preferred image. Later the decision was made to include videos as well. As is common with a lot of social media today, the videos were meant to autoplay without sound. In the middle of the software development, there was a major issue with playing two videos simultaneously. Such low-level issues commonly crop up in software projects and can delay the entire project. Either reserve some extra time for special features or be ready to postpone these features until later.

Further, a lot of time can be wasted by neglecting some of the basic issues surrounding the project. Common difficulties include

- New employees not being adequately equipped to start work

- Missing access rights

- Copying data from production to test and losing all the test data

- Slow development systems

Preparing and planning to avoid these kinds of hiccups requires having people on hand who are able to help swiftly. Otherwise, the time wasted quickly accumulates when no one takes responsibility.

In the next chapter, we'll run through the actual development stage. As well as giving you an outline of what this coding and engineering work entails, you'll also see how the input of the business side is still essential at certain stages in the process.

CHAPTER 14

Develop

After all this talk and planning, the development team can now finally get down to building some actual software. This chapter does not aim to explain how to program but rather gives an insight into the process—along with any surrounding tasks that can help the development team work as efficiently as possible. A good software developer should understand the nature of your business, and together you can ensure that you create the software to support it.

One of the central aspects of computer programming is the naming of things in the code itself. The terms used to describe the business problem are used again in this stage. As already described in the section "Ubiquitous Language" of Chapter 12, these terms need to be concise and clearly understood by everyone. Because the code models and represents real-life constructs, a sound naming convention will reap rewards. The company will not only benefit in the current project but even more so in future projects when a different set of people may work on the product and code. They'll find the clear definitions very helpful.

In many cases, a user will communicate with a system, but the communication can also be between two systems without human interaction. This could be a scheduled task or one caused by an event, such as when a share price reaches a certain threshold. Here, I will refer to "the user" for all these types of interaction.

© Jonathan Peter Crosby 2023
J. P. Crosby, *The Business Manager's Guide to Software Projects*,
https://doi.org/10.1007/978-1-4842-9231-0_14

At this point, the developers will need particular consideration as they knuckle down and start to write the code.

Getting Into the Flow

A programmer-friendly workspace allows the developer the concentration necessary for working well and efficiently. Psychologists call this ideal state *the flow*. Just as in the arts, this flow brings a kind of gentle euphoria as one largely forgets the passage of time.

This state of mind is essential for developers as well as other professionals such as engineers, designers, and writers who are occupied with *high-momentum tasks*. Work can only progress well when you're *in the flow*. Flow, at the same time, can't just be switched on and off. The developer, like the other professionals mentioned, has to concentrate for at least 15 minutes before reaching this state. Any noise or interruption can destroy the flow.

Not only is effective time lost through interruptions and disturbances, but the developer also suffers a great deal of frustration. Not all managers or colleagues realize the importance of this state of flow as most of their own work is done amid frequent interruptions. A working environment that is less than optimal can be another case of self-defeating cost-cutting. Insufficient quiet office space can result in low performance, and the cost of a developer's time ends up a lot higher than the space savings.

People should be free to take their work to another place where they're undisturbed by phone calls, desk visits, or general noise. If no quiet spaces are available in the office, the team could decide to redirect all calls to one developer and rotate this duty, thereby at least reducing the disturbances.

I was interested to learn from my cousin Ben of how his employer had introduced a 3-hour period for *deep work* every day. Ben, a project engineering manager, described this innovation as "simple and brutally effective." *Deep work*, a term coined by Cal Newport, refers to the undisturbed mental state that is necessary for working effectively on complex tasks—the tasks that can only be done in a place where there are no interruptions. In his book *Deep Work*, this computer science professor describes this work as "Professional activities performed in a state of distraction-free concentration that push your cognitive capabilities to their limit. These efforts create new value, improve your skill, and are hard to replicate." This means working in an environment without any talking, phones, meetings, Internet, or social media [Newport 2016].

Regular communication between all stakeholders nonetheless remains crucial throughout the development stage. In the next section, we'll look at the best ways to facilitate this.

Effective Communication

Here we'll look at the communication during the development phase. Regular interaction during this phase will help ensure the product is going in the right direction.

Most companies and teams I've worked for have used specific methods of communication during projects. Your teams can try out different ways of communicating and see which works best for them. Holding daily stand-up meetings to check progress can be useful, for example. Also, regularly performing reviews, both for the work completed and the work in progress, can prevent any big surprises toward the end of the project.

As already discussed, effective communication during the development phase requires a common language, but developers and domain experts sometimes seem to speak two different languages. In his book *Domain-Driven Design*, Eric Evans looks at the confusion that often results when domain experts give instructions to developers. One or two developers may succeed in making a sketchy translation, but their attempts to explain the information to the others are usually less than accurate. Both sides may use the same terms to mean quite different things. Any confusion makes for model concepts that are vague. The consequence of all this can be unreliable software where the parts don't fit together properly. To avoid such a negative outcome, the two parties need to work together in defining common terms and a common language. Once everyone starts to speak the same language, both sides can gain the knowledge necessary to work effectively together and achieve the full potential for the model. A domain model can act as the core of a common language for a software project [Evans 2004].

Don't get caught up for too long in arguments about things that have little effect on the project. I've witnessed such discussions too often in IT. If the reasons for project failures are usually sociological rather than technological, why do IT professionals spend so much time arguing about technology? Some IT professionals will completely dismiss a sound idea just because of one bad experience. In my opinion, such dismissiveness is linked more to a general fear of the unknown. Strong opinions need to be based on broad understanding rather than on such limited experiences. If the matter being discussed isn't really that relevant for the outcome of the project, then the team should aim to move on. In cases where the matter *is* critical to the outcome, the team will need to reach a compromise by discussing the reasoning in more detail. If this doesn't help, consider consulting a neutral subject matter expert or manager.

The project will also benefit from the teams agreeing on the *definition of done*. A piece of software is still not usually considered as completed when it is simply working satisfactorily on the developer's computer. There are actually many different definitions of "done"—it can mean that the software runs on the development server, automated testing has been done, the acceptance criteria have been met, or the software is running in production. Therefore, you need to agree on an exact definition within your project.

During the development phase, all the changes need to be proactively communicated to the testers, the maintenance team, and the producers of training materials. These stakeholders will need to adapt their work documentation, processes, and deliverables accordingly; otherwise, they may end up testing or verifying the original requests without the latest amendments.

Next, we're going to look at how to choose the tools and applications that will suit your company needs best.

The Right Tools for the Job

Basing strategic technological choices purely on the personal preferences of only one or two individuals is pure folly. Nonetheless, I've worked for various companies that have done just that—they kept changing their "strategic" programming language and "strategic" database system for all the wrong reasons. This kind of change usually coincided with the arrival of new upper IT management personnel or a new enterprise architect. These newcomers sometimes decided to impose their personal preference on the whole IT organization. I once attended a meeting where an IT manager explained that the programming language we currently used was like a family wagon, whereas the new one they were introducing was like a sports car. The comparison was completely false. Research has, in fact, shown for decades that the type of modern programming language used makes very little difference to the development effort. For most business systems, the choice of the underlying technologies and programming languages is far less important than some techies like to think. There are, however, a few very specific fields where certain programming languages are more suitable than others.

Do keep your eyes open for any relevant new technology, but at the same time, you don't need to keep up with the latest fashion of the season. If a company like Google can stay consistent with its preferred programming languages over the course of many years, then so can most other companies. I'm not saying that Google's chosen languages are the best choice for every business—just that you probably don't need to keep changing the strategic technologies every few years.

189

Many companies prefer to use and combine different technologies. Limiting oneself to a single standard type of technology restricts any experimenting, and in most cases, it's all about using the right tools for the task at hand. From a software development perspective, a core focus should be on the way that systems communicate.

Which are the most stable, user-friendly applications in your company? Can the strong teams behind these applications be scaled out and the knowledge shared?

These points may seem obvious, but I've witnessed a company replacing a good application just because a senior IT manager didn't like the programming language it was developed in. The team replacing the software didn't have a particularly good reputation for building solid tools, but they used the boss's favorite language. Maybe he used to program in that language himself. This replacement resulted in delays much longer than usual, important features that were missing, and, of course, unhappy users. The team was not particularly good at designing user-friendly software either and lacked knowledge of the business domain involved. Instead of focusing on how to add business value, the IT manager pushed his personal agenda and preferences—switching from one programming language to another for no technically valid reason.

Over-engineering often means that only a few people can understand how something works. The development team needs to keep the software as simple as possible.

Most applications that are built are nothing like Facebook, Twitter, Amazon, or Google. Tools that work for large Internet enterprises might not be the right choice for you. Some developers love to chase the latest fad just tweeted by their favorite developer hero. Such fad-driven developers may feel so enthusiastic to adopt a new technique they hardly consider the business context. Instead, developers should check whether a similar tool is already being used in the company. Other departments may have already solved the problem. Remember that introducing a brand-new technology may involve setting up a new support team for this technology within the company.

The following section will outline the use of patterns in software design as well as a few other minor topics relevant to this stage of the work.

Design Patterns and Other Best Practices

In this section, I've gathered a few miscellaneous topics relevant to the developing stage—non-techies will find it useful to know about them, and in some cases your contribution will be required. First, a word about design patterns. These patterns can be loosely compared with fonts, where you can portray the same letter in many different styles. According to the architect Christopher Alexander, "Each pattern describes a *problem* that occurs over and over again in our environment, and then describes *the core of the solution* to that problem, in such a way that you can use the solution a million times over, without ever doing it the same way twice" [Alexander et al. 1977]. Such patterns are available for solving many common problems in software programming too. The code patterns will have names like *factory pattern, state pattern, singleton pattern, strategy pattern*, and so forth. These patterns outline how a problem can be solved, regardless of the programming language or code structure used.

Most software projects are the sum of many, many simple mini-decisions. Only very rarely will a project involve a single element that is highly complicated, such as a long formula. The tricky part is in handling the organization, structure, and interdependence of all the smaller parts. The developer will aim to group and organize these parts in a way that makes it easy to understand, maintain, and extend.

One of the unpleasant tasks of the application support team is being on call and woken up at night (Figure 14-1). Luckily, I've never had to do that, but several colleagues of mine have—usually because badly developed software causes many false alerts. Besides being frustrating for the employees who need to get up at night, false alerts are also expensive, creating extra hours of work at night for the various teams affected and necessitating follow-up meetings and further analysis. As part of every software project, the team needs to decide on the criticality of issues in terms of the business as well as the technical side.

Figure 14-1. *For which issues does the on-call support need to be woken up at night?*

Let's say, for example, that data is delivered daily as a file. The file could contain bad or incomplete data; it could be empty, delivered late, not delivered at all, or sent twice. The way to deal with these problems will really depend on the type of data the file contains. It could be a simple product list that hardly ever changes—in which case the receiving system could possibly run for days using the old data. The difficulty would only become a real issue when a new product wasn't delivered.

The next item in this assortment of mini-topics is the *table-driven method*, one of the best ways to store complex and dynamic business rules. It is a very useful method for storing pricing structures or contract conditions such as insurance, mobile carrier, or legal conditions. The values can change over time, and different customers will have different deals. Storing this information in a table better facilitates retrieving, testing, and extending the business rules. Such a format can be easier for the business to verify than any rules hidden in code.

In the following section, we'll take a closer look at the user interface. As the part of the software the end user sees and interacts with, this area is of central importance both to the business and the software teams.

User Interface

The term *user interface* (UI) actually includes any device a user interacts with and uses to control a computer. The device could be a mouse, a keyboard, a lift button, a touchscreen, reading tools for the visually impaired, a classic screen, and so on. The abbreviation UI is also often used to refer to the *graphical* user interface, which is the visual design on the screen. For simplicity, I will also use UI to mean the visual screen design. The UI is one aspect of the UX (Figure 14-2), which stands for the whole *user experience* and the ease of use of the software. Closely related is the term *usability*, which establishes whether the user can accomplish their goal.

Figure 14-2. *UX vs. design*

Importantly, the UI is the part of a software system the user will see and interact with. Designers and/or developers create so-called *wireframes* and *mockups* to show what the UI will look like beforehand. The wireframe focuses on which elements go where on the screen, whereas the mockup is usually a more detailed and realistic version of the final design.

Getting continuous feedback from end users on the early UI designs helps the project team meet the users' expectations. The *structure* of the design, which accounts for about 70 percent of user satisfaction, is the way the various elements are organized. Can the user find them easily? Are related items grouped together, and is the design consistent? The *behavior*, which makes up around 20 percent of user satisfaction, relates to how it all works. Do the elements work and react as the user expects? The UI visual design presentation accounts for the final 10 percent of user satisfaction and concerns colors, shapes, text, fonts, images, etc.

A good way of getting valuable user feedback is to observe and possibly film the end user when they are first presented with an early functional version of the software. The user should be given very little information about the new software to determine how intuitive the system is. The feedback can help identify any issues overlooked by the specialists.

Default value is another important term often relevant to the UI that will necessitate a business decision. The term refers to the initial value set, which can usually be changed. One country in a list of countries could be preselected, for example.

Next, I'd like to explain the idea of software versioning, along with the reasons for sometimes having several versions of a piece of software.

Software Versioning

As the versioning of software isn't globally standardized, there are many different numbering conventions. One of the most common ones is sequence-based, as shown in the preceding software version numbering (Figure 14-3).

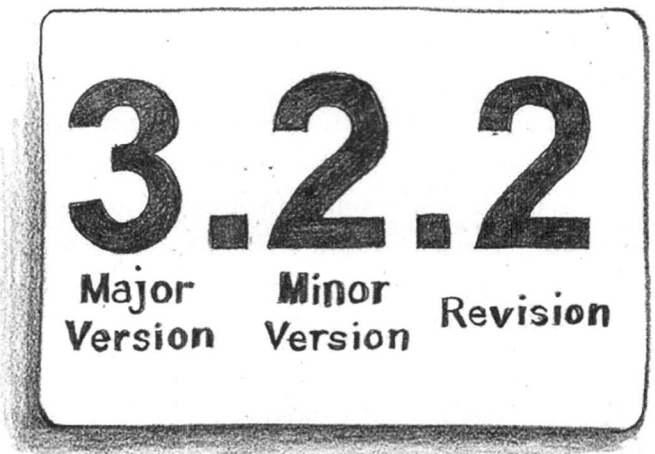

Figure 14-3. *A typical style of software versioning*

For commercial software, numbers below 1.0, such as 0.9, usually indicate a beta version that's meant for testing and not for official release. The change in the *revision numbers* of software will typically include smaller fixes, whereas a *minor version* may additionally contain some new features, and a *major version* could include larger changes. The minor version upgrades are most often included in the license, and further costs are tied to the major version increments.

A new major version could include *breaking changes*. In simple terms, this means the way to communicate with the software has been changed, forcing other systems to cater for these changes on their side too.

The next item on the agenda is that mysterious activity at the core of every software project—the process of coding.

Coding

Coding is the part developers usually enjoy doing most and is the reason they chose the profession—many express the wish to do a lot more coding.

It can be a fun challenge, like solving a riddle, a puzzle, or a sudoku. When coding, a developer can experience many small challenges, frustrations, and moments of success even before the first coffee break.

The various facets of programming described in this section will help you understand some of the terms developers use most often in their daily work. These terms are also relevant for communication between the business and technical teams. The business may require data extracts from the system, for instance, and may need to understand the terms used to describe the data.

Code is often written according to internal guidelines and standards. Research shows that developers spend around 5 percent of the time writing code and spend the other 95 percent reading it. Therefore, it makes no sense to speed up the writing process by 20 percent and end up with illegible code. It's easy to write code a computer can understand, but much harder to write it so that other humans can follow.

It's good practice to use names that are clear and meaningful in the code because sooner or later these terms will be discussed between the developers and the business. It's best to avoid generic terms like "GetData" when "GetCustomerName" may describe what the code does more precisely. The use of positive rather than negative labels can also improve clarity. Take, for example, a simple Yes/No field in a car rental system asking if the car has a sunroof or not. If the code label to store this data is named "VehicleHasNoSunroof," then "No" would make a double negative and would actually mean "Yes, the car has a sunroof." Such labels can be very confusing, especially with more complex business terms. Unfortunately, many systems contain this type of naming issue. The same is true for comparison fields like "AccountBalanceIsLowerThanZero." In this case "No" means the balance is positive. It's much clearer to indicate Yes/No for "VehicleHasSunroof " and AccountBalanceIsZeroOrAbove."

Will anyone understand the code that's written today a year from now?

If a developer says "I've checked-in" while at work, it probably has no connection with an online check-in for a flight or hotel. Rather, *code check-in* means to combine an individual developer's code changes with the team version. A brief description of the work done should always accompany the changes. The developer could have changed something that breaks something else accidentally. Being able to track who changed which code and why is extremely valuable in solving such issues. Also, automated code tests can be helpful to ascertain whether anything existing has been broken and no longer works as expected.

Code reviews and *pair programming* are techniques used to improve the quality of software. Code reviews are done before any developer's code changes are checked in. Another developer reviews the code, and the reviewer's name is documented. Personally, I've found that reviewing code with another developer is highly beneficial in terms of improving quality and sharing knowledge. Pair programming, as the term suggests, refers to two programmers working together on one computer. One person writes the code, while the other observes, and the two frequently switch roles. These techniques help reduce defects and involve sharing detailed knowledge of the written code, but do take more time.

Now for a bit more of the terminology you may often hear during an Agile project. *Continuous integration* refers to the process of each developer making small changes and running automated tests to make sure nothing has broken or to fix it quickly if it has. *Continuous deployment* is the process of software being automatically pushed to the production system once it has passed all the tests. *Continuous delivery* is an extension of this process and allows the same changes to be pushed to production, but in this case the business teams decide when a change becomes productive by switching the new feature on.

The developer will run the program to check how changes have affected the software. Many programming languages include a process called *compiling* as part of running software—this converts the code from a human-friendly language into a computer-specific one. Depending on the size of the software program and the speed of the hardware, the process of compiling can take some time (Figure 14-4). Having a device that compiles quickly can strongly increase both productivity and job satisfaction.

Figure 14-4. *Compiling can be a lengthy process, especially on inadequate hardware*

Software uses *logging* to track many of the stages when an application is running. The development and support teams will typically define the items they want to see in the output, the log file. The support teams can use this file to resolve bugs. Some applications require an audit log, and logging is also used for security purposes. For example, logs typically track the details of financial transactions in chronological order. The quality of the log file will usually improve over time when a system becomes more mature, following real support cases.

Additionally, the philosophy of an open-failure culture is especially beneficial in software development, as discussed in the "Fail Fast" section of Chapter 12. Jim Shore describes this nicely, "When a problem occurs, it fails immediately and visibly. Failing fast is a non-intuitive technique: 'failing immediately and visibly' sounds like it would make your software more fragile, but it actually makes it more robust. Bugs are easier to find and fix, so fewer go into production" [Shore 2004]. For developers, there's often a Chinese wall hiding them from the production system. In many cases they are unable to access or view the production environment and hence don't even experience the same performance issues as users. To address this blind spot, developers should be provided with data very similar to that of the production environment. If confidential, the data should be anonymized. This data will allow the developers to identify performance issues in the development system as well as help them find issues in the user interface—where the data displayed is too long for a field, for example.

The development team is responsible for reporting back any changes in logic that differ from the original requirements to the project managers and requirements engineers. These project members will then update the documentation and the domain model. If this task is neglected, the language of the domain and code will no longer be properly aligned. In a system I inherited, for example, a field had been renamed from "order giver" to "order taker" in the user interface. The code and database had not been updated, probably as an attempt to save costs. Since the original team members had all left, it took some time to retrace this change. Fixing this issue generated much higher costs than simply aligning the names in the first place would have done. Allowed to accumulate over time, the many such smaller changes that haven't been aligned will cause major headaches. These could result in a system slowly turning into a legacy system that needs a complete (and expensive) replacement.

Refactoring code means changing the internal design of software without changing its functionality. You could picture it as a bridge over a river. The driver of a car does not need to know how many pillars are holding up the bridge, so long as the bridge is stable and safe to drive across. In theory at least, the pillars could be exchanged as part of the maintenance work in such a way that drivers don't even notice. The reason for refactoring code may be due to a deeper understanding of the business domain or to make the software faster or easier to extend.

To conclude this section on coding, I'll leave you with a quote from Robert (Uncle Bob) C. Martin that's directed at programmers: "Always leave the code a little better than you found it" [Martin 2014].

We can now move on to explore the area of testing. Designed to check different aspects of the software, the various types of tests conducted throughout the project are vital for ensuring quality.

CHAPTER 15

Test

You'll find that every stage of a software development project includes testing. A quick online search using the term *software testing* will bring up over a hundred different types of tests. This chapter will list the most common types and mention the ones most relevant to the business departments. Testing and requirements gathering belong to the most important business tasks and, as such, contribute substantially to the success of the project.

Types of Tests

You can expect to be closely involved in the *user acceptance tests* (UATs), and we'll look at these first. The order of the tests listed here is no indication of the order in which they are performed.

User Acceptance Tests (UATs)

The UATs are typically among the final tests done before the new system or the changes are taken into production. As the name suggests, the tests are conducted by end users on a pre-production/test system. Business representatives, business analysts, and dedicated testers are usually involved in these tests too. As the system should have reached a stable state by this stage, the go/no-go decision will depend on these final tests.

© Jonathan Peter Crosby 2023
J. P. Crosby, *The Business Manager's Guide to Software Projects*,
https://doi.org/10.1007/978-1-4842-9231-0_15

The UATs are based largely on test cases and on the acceptance criteria defined in the requirements. A set of acceptance criteria should be attached to each requirement. The burning question here is: does the software meet the criteria specified in the requirements? The outcome of these test cases will probably need to be tracked as test evidence.

To ensure a high level of reliability in the results of these tests, the UAT and production systems need to be near-identical. If there is no test environment at all, your teams need to build one up before even thinking about going live. The infrastructure, code, and most of the configuration must be identical. Any differences between these two environments should first be resolved. Both the type and quantity of test data need to be very similar to the (expected) production data; otherwise, the tests cannot ensure that the system will work in production.

The UAT system needs to run stably, be available, and have the necessary resources to support it. Although the support priority for the production environment will always be higher than for any test system, UAT support is still important. The project's progress may in fact depend on it. The attitude "It's just a test system" can be detrimental if any surrounding systems depend on receiving data or vice versa. In one extreme case with a client of mine, a project was substantially delayed due to the bad state of a connected UAT system.

Unit Tests

One of the smallest elements of testing is called a *unit test*. These tests check very small parts of the software code, such as the depositing of money in a bank account. In a simplified example, let's say a customer deposits $50 into an account already containing $800. The unit test checks if the system correctly adds up these figures to $850. A system may have hundreds or even thousands of such unit tests. This type of test is typically created by the software developers themselves as part of coding. During the development phase, the tests are run throughout the day to ensure nothing has been changed that shouldn't have been. Some programmers even write the tests before writing the code that does the work. This approach allows the programmer to think through what the code should and shouldn't do in advance.

The *test coverage* is a term used to express the percentage of the code that is being tested.

Integration Tests

Built upon the unit tests, the integration tests are created by developers to help ensure that the separate modules in the software all work together. There could be a module for communicating with the database and a further module containing the business logic. A third would be responsible for sending user interface content to the user's screen. An integration test will simulate the passing of data between modules, for example. Both the unit and integration tests are typically run on the developer's own computer.

System Tests

The system tests that follow the integration tests involve running the code on a shared system such as a central development server. The developers create these tests too.

Regression Tests

The regression tests are used to make sure the software, especially the UI, is still working as before and that nothing has changed that shouldn't have. These tests can be run by automated test software and/or performed manually. On larger projects, there'll usually be a dedicated test person or team conducting this work. The regression tests are more relevant for existing systems than for new ones. Sometimes these tests are also performed in the early stages of development to assist the developers in identifying unwanted changes. These test cases are a team effort and can be defined by anyone from the project team, not just the developers.

Penetration Tests

Also known as *pen tests*, these search for any vulnerabilities in the system and are sometimes executed by an external company. This company will simulate an attack, try to compromise the system, and report the findings, thereby revealing the security weaknesses of the system. Such tests, which can either be automated or manually executed, are typically conducted in a test environment. Different companies may perform these tests from time to time to provide differentiated views of the system.

Smoke Tests

This test is named after one done by circuit designers when they switch on the electricity to test a circuit board. If no smoke appears, they expect the circuit will work. The software smoke test simply involves starting up the application and doing a few clicks to check that the application responds as it should.

Load and Performance Tests

Load and performance testing creates a simulation of concurrent user access to the system and shows how well the software copes under pressure. The measurements can indicate if the software is too slow when the expected number of people is using it. The simulated load can be gradually increased to see which component in the chain fails first and is hence the bottleneck.

Business Readiness Tests

These tests are the final ones conducted just before a system is released to the end users. At this stage, the software has been installed on the production servers, and everything should be ready to go live. The tests are done on the live system. The business departments or representatives who conduct these tests will give the final go or no-go decision once these tests have been completed. If any big issues come up that lead to a no-go decision, the new software is removed and the old version restored in a process known as a *rollback*. Otherwise, in the case of completely new software, the go-live is simply postponed.

A/B Tests

The A/B tests are used in a live system to compare different scenarios where a limited number of users have a new feature turned on, while the other users cannot yet see the new feature. This type of testing can be used to check whether the feature is achieving its goal and running stably before making it available to everyone.

Test Automation

Test automation replaces the manual regression tests normally done by a test person. The test automation software will typically emulate a real user and the way they would use an application. Test automation goes even further and can also test system-to-system interaction, database communication, and many other areas that might be difficult for a test person to perform. This software can run many more test cases than a manual user could in the same length of time.

Automatic regression testing can be very challenging, especially where there are many moving parts. Most changes to your application need to be updated in the test automation tool—so the automation is an ongoing commitment. Automated regression testing is a powerful tool when implemented correctly though. The software to be tested usually needs to be programmed in a way that caters for automated testing. This preparation could, for example, enable the automation software to uniquely identify screen elements such as buttons and text boxes.

Test automation software can give an overview of test coverage and test status throughout the project. The test data can usually be managed within the automation tool itself or can be stored in a database or elsewhere. Test automation is also widely used alongside Agile work methods to achieve continuous delivery. In 2017, some companies reported that over 90 percent of their tests were automated.

I enjoy working in environments where test automation is set up expertly—it results in fewer bugs and more stable software. However, I believe we still need manual testing to supplement the automated tests. The human tester can notice issues that a computer doesn't detect or that are too time-consuming to replicate. At the same time, of course, the automated testing software is continuously being advanced to cover new areas.

Test Data

High-quality test data is crucial to the outcome of the testing results and directly impacts the quality assurance of the final product. Testing an application with poor test data can give a false sense of security, as the tests will not have covered all realistic and common scenarios. Test data preparation is, however, often treated as an afterthought nonetheless.

Test data may need coordinating across multiple systems and will usually be linked using a unique identifier (ID). Customer data, for example, may be available in three different systems: a CRM, an order system, and a document archive. Each test system should use the same ID in its test data to identify the same customer; otherwise, the testing across these three systems will fail. Because a change in one system's test data can have an impact on the other systems, cross-system alignment of test data changes must be done on all sides.

Additionally, the test data will need regular refreshes. Data updates are often necessary when products and services reach expiration dates; otherwise, this factor may prevent some features from working. An event booking system could display events that take place in the upcoming month, for instance, and these events will quickly expire. The main screen, instead of showing upcoming events, would be empty, and the search may bring no results if no refresh is done. One solution is to regularly copy the data from production into the test environment. Another option is to regularly and automatically change the dates of these events to a date in the near future. Refreshing the data needs to be well-coordinated; otherwise, the open test cases may no longer be valid.

During the process of copying data from production to the test system, any confidential data, such as the salaries in a human resources system, must be anonymized. Another important step in this copying procedure is restoring any overwritten test users' access rights. Neglecting this step may land you with the common problem of blocking a lot of people from working.

Attention and even controls are required to ensure that no test data finds its way into a production system. An example involving the US stock exchange Nasdaq illustrates the kind of thing that can go wrong. A data glitch had allowed test data to enter the live system, causing a rather loud hiccup in the stock market. It led to incorrect prices and indicated, for example, that Amazon had suffered massive losses in market capitalization and that Zynga's share price had soared by some 3,000 percent [Wigglesworth, Hughes & Lockett 2017]. It is, in fact, technically possible to set up systems in a way that prevents any test system from writing data into a live production system.

Further Aspects of Testing

A risk-based approach should be used for testing. This involves classifying each test case according to *damage* and *frequency*. The business teams assess what damage to the company would result if a certain part of the software failed. The frequency refers to how often the same part of software is used—existing systems should already have such statistics, and for a new piece of software, the frequency is estimated. The damage and its frequency are both assigned an exponential number such as 2, 4, 8, 16, etc. accordingly, where a low number represents low damage or low frequency. These two numbers are then multiplied to indicate which failed test cases have the strongest impact. A credit card system failing to process payments in stores, for instance, would have a high frequency and would cause serious financial damage to the company as well. The multiplication of these high figures would show a very high risk for any problem with this part of the software. Therefore, a great emphasis should be given to testing this software feature. There will also be many other defects with a much lower priority. Knowing where to focus the effort by referring to these classifications can make a big difference in testing.

Testing also includes *corner*, *edge*, and *boundary cases* that focus on ensuring the limits work correctly. The corner case tests a value that only just falls within the allowed range of values. The edge case tests the minimum or maximum value, and the boundary case tests a value beyond the minimum or maximum limits. For example, when someone applies for credit online, the maximum credit might be $10,000. The corner case would test $9,999, the edge case exactly $10,000, and the boundary case $10,001. This testing is not just for number ranges but can also be used for a minimum or maximum number of allowed characters or other criteria.

If the software displays many different languages, the testing will check that texts and text blocks are properly formatted for all the languages used. This testing will apply to the user interface as well as to customer letters and any other output. The word "Subscribe" may fit well on a button, for instance, but the Icelandic equivalent may be longer and hence too wide.

When an old system is being replaced with a new one, certain financial reports or other outputs may need to stay the same as before. Any differences will need to be explained. The replacement project may uncover issues with the old system. On the other hand, the new system may also contain mistakes, or both systems might be calculating something incorrectly. This type of testing, which involves analyzing and aligning two systems, is a very time-consuming but essential task.

In the following chapter, we'll go over the final tasks that need doing before the go-live.

CHAPTER 16

Training, Going Live, and Maintenance

The next steps lead up to the go-live of the new system or the new features. How the launch is done, again depends on the methodology used. Continuous updates may be used for an Agile process, and this more gradual approach can soften the landing. On the other hand, the Waterfall methodology will typically involve a large big bang release.

Irrespective of the methodology used, for every piece of software, there'll be an initial version that requires user training. After working for several months on the software, the project team will know the system like the back of their hands—but the new users may still know nothing about the upcoming release. One of the project tasks is now to close this gap, even if the system is for internal use only.

Besides, upon go-live, any connections to other systems will now be activated, and the maintenance team should be ready to begin productive support.

The rest of this chapter will look into these tasks that ensure a smooth start for the new software. By carefully following these steps for your project, you'll reap the rewards of all your efforts in collaborating so well as a project team.

© Jonathan Peter Crosby 2023
J. P. Crosby, *The Business Manager's Guide to Software Projects*,
https://doi.org/10.1007/978-1-4842-9231-0_16

User Training

Preparing and coordinating good user training for the software will greatly increase user acceptance. Your company may have training facilities to prepare the users. Further options include live online sessions, recorded training videos, and/or training booklets.

The setup and maintenance of an effective training system can involve some time and planning. If the system to be taught connects to other systems, these connections may need simulating; otherwise, the training system will become highly dependent on the other system's availability—possibly interfering with a training session.

When showing realistic scenarios, trainers need to use training data that is similar to real data while avoiding the use of any real data—especially if the system contains confidential information. Test data can expire, and the same is true for education data. Searches for events, bookings, tickets, or other date-driven data may no longer work if no search results can be returned.

For one of my clients a few years ago, we organized an internal user training session for an important piece of new software. Users had to attend a compulsory training session before being given access to the system. After completing the 2-hour session, the trainer sent the list of attendees to the administrator to activate their accounts. I'm convinced this useful and compact training helped the software reach swift acceptance—and with far less support than in other comparable software launches.

Support and Maintenance

Well-prepared technical and business teams should be on hand to assist users. The software will typically include the admin features required to perform this support work.

Do all levels of the support organization know how the user management will work? What might be the most common support enquiries?

For completely new software, the maintenance teams will have gained some hands-on experience already by supporting the test environment. Most pieces of software going live encounter some teething problems—handling these well greatly increases user acceptance. Showing new users that the software support is swift and efficient will help create a positive momentum for all new users.

Whenever a valid bug is reported to the support team, the development and testing teams need to be involved. Beyond solving the problem, consider also adding unit, system, and regression tests, especially for critical bugs—these prevent the same issue from recurring. Further, any regular tasks that need to be executed by the support teams should be well-documented—particularly if a task is required only once every 6 months or so. Otherwise, the rare tasks are easily forgotten, especially in future handovers.

A good system design makes the system transparent and easy to support. This may sound like an obvious statement, but many of the systems built are incredibly difficult to support, and issues can be hard to resolve. The support teams may not be able to view all parts of the system they support, for example. In a few cases, there can be areas of an application, such as employees' salaries, that even a trained support specialist shouldn't see. In my experience, however, most nontransparent areas are not due to such required restrictions but are simply down to poorly designed software.

By providing a Frequently Asked Questions page, hosting an online discussion area, and adding some self-service features, you'll greatly reduce the amount of support work needed. Users can then perform common tasks like resetting passwords, ordering services, and editing their profile without the assistance of a support person.

A Service-Level Agreement (SLA) is a type of contract used to define the support team's commitments in areas such as availability, quality, service hours, and responsibilities. Normally, this contract will be drawn up and will go into effect once the first version of the software goes live.

Monitoring and Analytics

Considerable benefits come from providing the business and IT support teams with adequate monitoring tools. These can give real-time information on the health of the application as well as on user satisfaction. The tools monitor performance, availability, errors, user behavior, and network issues, along with many other aspects of the application. Further, these monitoring tools allow the business teams to observe the countries that users are connecting from, whether a new feature is being used, reasons for the sales on their website to suddenly rise or fall, conversion rates, and a host of other things. The *conversion rate* refers to the percentage of visitors who make a purchase on a website or app.

These monitoring tools often include user interfaces that allow support teams to actively observe the systems. Additionally, the tools can generate alerts once any predefined condition is met. The tool will respond if, for example, the speed of the system suddenly drops well below the average of the previous 7 days. It could generate and send an email, text message, or support ticket to the support team. Having the right initial setup to monitor the application is therefore well worth the investment.

Synthetic monitoring is used to replicate the exact behavior of a user in a live environment—it can further help in monitoring the system's health. By imitating a customer, for example, the online shopping experience can be tested every 5 minutes. The monitoring tools would then measure performance, availability, etc. as though a real user were accessing the system. The fictional user will not be able to place real orders, and the actions made should not be included in the real user statistics.

The analytics of a system are measurements made on the system to gather usage and other statistics. The patterns gathered from this data help teams to interpret user behavior and to perform a deeper historic analysis. Knowing and understanding how a user navigates through your software can greatly assist in improving the business performance. One visual representation often produced using analytics is a heat map—this shows how much the different areas of a web page or screen are used. The frequently clicked areas are marked red (hot), and those less frequently clicked are shown as blue (cold).

Through which link did a user access our web page? How often is this feature used? What other products may the customer also be interested in?

Many business teams benefit from utilizing live monitoring and analytics. It's important for you to start considering these two distinct topics in advance. One of your tasks will be to define the most relevant information for your later analysis.

Now that all the preparation and development work is finally done, the software can go live.

Going Live

All the preparation, the many big and small decisions, the hours of overtime worked, and all the worries have led up to this point. The moment has finally arrived when the wider audience can access the new or changed system. Only now will the project teams know for sure how well the system works, whether the end users find it easy to use and whether it meets their objectives.

As mentioned, the go-live can be done in one go with the *big bang approach*, where all parts of the new system are available from day one, or, alternatively, using the *phased approach*. In this case, the different groups of users are given access gradually. If the new system is replacing an existing one, there could be a parallel go-live—certain features would be disabled in the old system and replaced in the new one.

The so-called *release notes* are published with the go-live. These notes will contain a list of all the features included in the initial version or the changes added since the last version. Typically, the release notes will contain very brief descriptions of the main features and of the major bugs fixed.

The code, which must be identical to that last approved in the test environment, needs to be installed or copied into the live environment. Code that's not completely identical can be a new cause of errors that are very hard to identify. As part of the go-live, a data migration may be required—involving copying data from an existing system to the new system. If a substantial amount of data needs to be copied, it could mean many hours of copying before the new system can go live.

Installing the latest version on the live servers is often done in the evening or at the weekend to cause less disruption for users. Once the system is ready, the business representatives will run a business readiness test, and if this is successful, the software will go live. As end users will inevitably require some support at the launch of the new software, the technical and business support teams can help provide this. Where possible, floorwalking can be useful for answering any urgent questions the end users may have.

Congratulations! You've done it! It's time to celebrate and then, at some point, to reflect on the lessons learned throughout the whole project.

Pitfalls—Part 2

One of the crucial factors in the final phase of the project is how well the various IT teams, including the development, testing, operations, and support teams, work together. Each IT team will have its own horror story of working with other IT teams: of how everything went wrong with things like poorly tested code, badly designed databases, or incorrectly configured servers. Such standoffs between development and IT operations teams, for example, shouldn't be left to descend into tribal warfare. Just as you cannot achieve goals when the business and IT teams work against each other, the same is true for the IT teams themselves. The focus always needs to be on the final business goals, and the steps along the way must be based on sound collaboration. This cross-teamwork should lead to continuously improving and automating the processes of releasing software successfully.

The best processes are practically seamless—where you can hardly tell that so many teams and automated background processes are involved. Sometimes though, I've known the reverse to be true. Some teams can function as unnecessary gatekeepers instead of acting as enablers or supports. Keen to make everyone aware of their presence and importance, they fail to add any real business value. Such gatekeepers really need to work much more closely with the business departments. They need to understand the company's business goals and reserve the gatekeeper function for when something really is wrong. Breaking out of this mode and seeing the issue from a user's perspective can really help in finding a swift solution.

PART III

Technical Guide

CHAPTER 17

The Technical Side

Part 3, "Technical Guide," has been written for those of you who'd like to dig a little deeper into the more technical aspects of software development.

Extending on the initial conceptual guide, I'll introduce some further metaphors and analogies to help make everything clear. We'll look a bit more at how code is used to build software, touching on a range of other topics that include the security of your website; scalability, which refers to the capacity of the software and the number of users it can handle; as well as a section that explains the meaning of the *state* of software. Because the depth of technical detail you'll require for different projects may vary, you can choose the ones you feel are relevant to you. If you have no need for this technical information, you can simply skip this part and move on to the final words in Chapter 21. You might also like to enhance your knowledge further by reading through some of the less technical topics in "Appendix C: References and Further Reading."

We'll start by outlining a couple of points relevant to the planning stage.

© Jonathan Peter Crosby 2023
J. P. Crosby, *The Business Manager's Guide to Software Projects*,
https://doi.org/10.1007/978-1-4842-9231-0_17

CHAPTER 18

Coding and Design

Theme Parks, Jenga, and More— Structuring Code

The actual code can be organized in many ways. You can imagine that most programs start off just like a blank sheet of paper and the development team decides what to put where. They can determine how they want to name, structure, organize, and group the code. Although there are many patterns, best practices, and recommendations, each program is customized quite differently. This customization usually involves organizing the code into higher-level concepts. One section of code will be responsible for the *customer details*, for example, another for the *purchase order*, and so forth.

If you were responsible for planning a new theme park (Figure 18-1), for instance, how would you arrange the layout of the different rides? Would you group the ones for the youngest kids together, placing the fast rides for teenagers in a separate section? Would that suit families and groups with kids of different ages? The older kids would be bored waiting around while the younger ones were playing and vice versa. Should you mix the rides instead? Should there be a large central food court or small food outlets spread throughout the park? This example may seem a bit far-fetched, but it illustrates the similar choices the developers need to make when planning software. Which parts do we group together, and which

© Jonathan Peter Crosby 2023
J. P. Crosby, *The Business Manager's Guide to Software Projects*,
https://doi.org/10.1007/978-1-4842-9231-0_18

elements are required everywhere? As you can imagine, such decisions are not always easy to make.

Figure 18-1.   *Laying out the parts that belong together is an integral task in software development*

Another example of organizing code can be compared to designing a closet. Apart from reserving space for hanging certain items, you could also make some small compartments and shelves to organize everything else. Would you arrange the items according to the type of clothing, the color, or the fabric? Too much separation might be tedious, but insufficient shelving would mean the opposite. Piling up lots of items together would make it messy and disorganized. Similarly, coders need to strike the right balance for the size of code chunks (Figure 18-2). Code can also be separated into many small parts. In general, the programmer should avoid making big parts that include too many items. It's best practice to make a small part of code responsible for performing just one task.

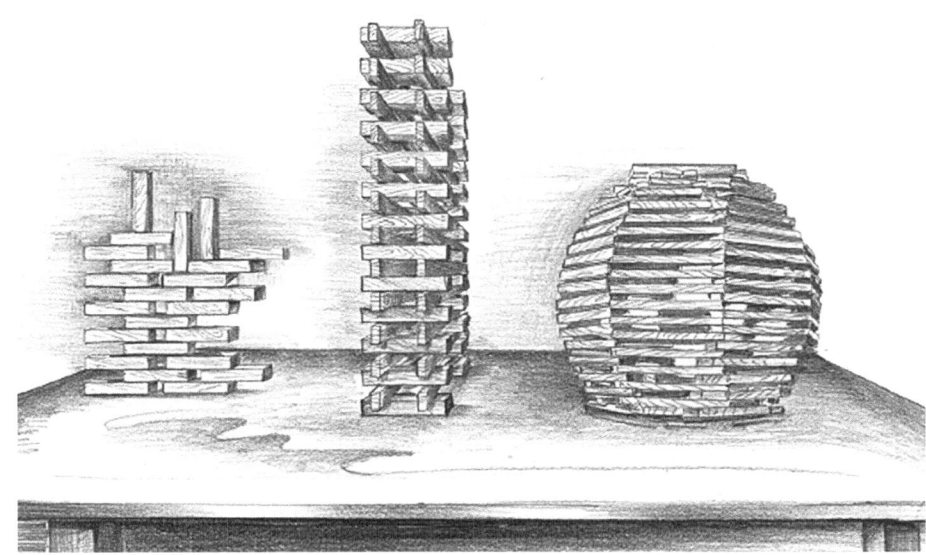

Figure 18-2. *Ideally, software is built using as few pieces as possible*

A simplified analogy of building good software is like building with Jenga blocks. The goal should be to use as few pieces as possible without undermining stability or sacrificing shape or function. This recommended minimalistic style also means that just counting the lines of code reveals very little about the quality or how far the development team has progressed.

It's important to understand that code from two separate programs cannot easily be merged, even if created by the same software company. That's because the internals of code are usually quite different from one piece of code to another. This situation can also be true for packaged software, especially if the software product has been customized to a company's business needs. Non-technical members of management are often unaware of this factor. Understandably, they assume that the products and data will be easy to combine and will be fully compatible.

The next topic in this chapter is system interfaces. When two systems are connected, data flows from one to the other through a system interface. For a business system to function well, data needs to flow smoothly. The teams need to ensure the interface functions efficiently to avoid all kinds of hiccups and delays when using the applications.

System Interfaces

As mentioned at the beginning, the inspiration for this book came when Daniel asked me, "What exactly is an interface?" In answering his question, I referred to the underground pipes that connect to a house. "If you build a new house, you'll probably have pipes connected to the property—some to supply water and others for drainage purposes."

Similarly, a system interface is used to send and receive data to and from another system. You may be building a payments system, for example, and need to access customer addresses to send out bills. The customer data is stored in a specific customer system. The payments system will request the address data required using a system interface. Typically, the requestor will send a unique identifier (an ID), and if authorized, the data will be retrieved.

Imagine you're converting a bedroom into a bathroom. If capped water pipes already exist in the room, the work will be fairly simple. On the other hand, having to install new pipes that lead to the central water source will involve much more work. In software projects, a lot of the work revolves around having the right data in the right place. For a new feature in an online shop system, for example, customers may want to see the guarantee details on their order history. Is this data already accessible in the shop system, or should a new "pipe" be plumbed to the product system to make this data available for customers on that page? Having to extend the interface for this additional data will involve much more work.

An interface can also be understood as something like a power socket in a wall. Different types of electronic devices can be attached to the socket. Likewise, with interfaces, the so-called *signature* lays out the format to send and receive data. Interfaces also allow for interchangeability; the connected systems can be replaced without one affecting the other. A microwave can be unplugged and a phone charger inserted—you wouldn't need to make any adjustments to your power socket.

Although most business systems require an interface, not every piece of software needs one. Just as a log cabin in a remote location can be habitable without running water or electricity, certain independent software applications don't need any interfaces at all.

Just as your car battery probably couldn't power a large refrigerator, a system interface will only be able to process a certain amount of data. That means the person responsible needs to identify and specify the required capacity in advance. Too much data being sent back and forth across a system can cause delays.

In the following pages, you'll learn about *scalability*—the capability of the software to handle a surge in usage.

Scalability in Software Projects

In business, *scalability* traditionally indicates a company's ability to increase its profits at a higher rate than the costs rise. Scalability in teams refers to a team's ability to handle growth, either by improving its processes or by enlarging the team. In IT, you'll find that scalable software is designed to easily handle an increase (or decrease) in access and performance without disruption (Figure 18-3).

Note   If you send out a marketing email to 5,000 or 10,000 people, can your website handle the load if 3 percent of the recipients click the links at around the same time?

Figure 18-3. *How well can the lifts cope in the rush hour?*

Scalable IT Systems

When lots of people use a website at the same time, the users' devices could be connected to one of many web servers, although the website visitor will be unaware of it. This practice of utilizing multiple servers simultaneously is known as *load balancing*—it's like having multiple elevators in a building. If the world's tallest building had only one elevator, people would certainly lose a lot of time waiting. An efficient elevator system can cope well with peaks in usage. In some buildings, certain elevators serve specific functions like transporting goods or stopping only at special-access floors. For this example though, imagine a building where all the elevators function identically. The web servers for one software program would all be identical, too—a visitor to the website expects the same experience regardless of the server they are connected to. If more capacity is required at some point, the team can either *scale up* or *scale out* the servers.

Scale-Up

If the capacity of the elevators needs to be increased, you could speed up the existing elevators by using more powerful motors. In IT, scale-up refers to making the servers run faster by adding more powerful components, such as additional memory or faster processors. This method can be a relatively quick and inexpensive way of achieving higher performance for an application—purchasing computer hardware is often far less costly than bringing in a skilled workforce to solve performance issues. However, once the scale-up limits are reached, other options need to be considered.

Scale-Out

Scale-out means adding more servers, just like adding more elevators. The process is much easier if the development team has already programmed the application for scale-out—as is common practice these days. Adding extra servers would mean the web page or application can connect to one of numerous servers located in one or more data centers. If one of the data centers is offline for any reason, other data centers would serve the users instead. For systems that are used globally, these data centers are spread out around the world—the closer the server is to the users, the faster it will be under typical conditions.

Scale-Down and Scale-In

If you want to reduce the size of the IT infrastructure to save costs, you can do the opposite of scale-up or scale-out by scaling down or scaling in.

Reasons for Scaling

These processes of adding or reducing the number of servers can even be set up to function automatically. Faster or slower servers can be switched automatically too. These options help to regulate the costs for peak times and other periods of lower activity. The reasons for scaling up or out would normally be

- *More people* accessing the service

- The same number of users requesting *more data* from the service

- The same number of users requesting the service *more frequently*

- *Other IT systems* requesting more services or more data

- A combination of the preceding reasons

Some performance problems cannot be solved through scale-up or scale-out, however. Badly written software cannot really be improved much by adding more and/or faster hardware. Too much duplicate or unneeded data being sent between devices could be the root cause of these performance issues.

Bottlenecks

Imagine a cinema complex where the ladies' restroom has two toilets, seven hand basins, and seven hand dryers. There would be a long queue for the toilets, while the hand basins and hand dryers would not be fully utilized. The cause of the bottleneck would clearly be the shortage of toilets. Adding further hand dryers would not help reduce the queues. Any improvements should address the true cause of the bottleneck.

Similarly, for software systems, you need to identify the bottleneck that's slowing down the system. It could be a hardware problem, but it's more likely to be software related. Tools are available to help you find the cause.

"Eliyahu M. Goldratt, who created the Theory of Constraints, showed us how any improvements made anywhere besides the bottleneck are an illusion. Astonishing, but true! Any improvement made after the bottleneck is useless because it will always remain starved, waiting for work from the bottleneck. And any improvements made before the bottleneck merely result in more inventory piling up at the bottleneck" [Kim, Behr & Spafford 2013].

The next topic is on the *state* of software. This term refers to how the system handles the different parts of the software working in parallel and remembering what every part is doing.

Sports Stadium—Software State

The *state* of software is a concept that can be a little hard to grasp at first. Incidentally, it can also be a great challenge for software developers to get right. Picture the scene in a stadium just a few hours before a big football match begins. You can probably imagine many different activities going on at the same time. Staff at the food courts are preparing to serve customers, the teams are warming up, the security personnel are doing the routine checks, and the groundskeeper has just finished mowing the grass. Imagine how the groundskeeper would feel, returning from a coffee break, to find that the grass had magically grown back. What sounds impossible in real life can be a painful reality in software, however. I'm sure that most people have experienced losing the data they had entered on a web page or lost some of their work. The loss of data can sometimes be caused by poor state management in software.

Figure 18-4. *In a sports stadium, many independent activities go on at the same time*

In the stadium building, meanwhile, employees are using various devices and machines related to their respective work (Figure 18-4). Everyone is concerned with their own task. The caretaker is in an elevator on the way to repair a broken lamp, while the chefs are busy in the restaurant kitchen. Most of the amenities within the infrastructure operate independently. The elevator has no information on how many ovens are being used in the kitchen, for example, neither do the ovens store data on which direction the elevator is moving.

Similarly, in most software, many individual parts perform tasks independently of the others. The state of the software program is defined by the different active users and connections currently accessing the system and causing some form of processing. The system will support many different states at the same time. What happens if you lose your

Internet connection just after you've added an item to your shopping cart in an online shop? Will the software remember your state when you return to the shop the following day to continue with your order?

A typical diagram used to depict how the system internally handles state is called a *state machine* (Figure 18-5).

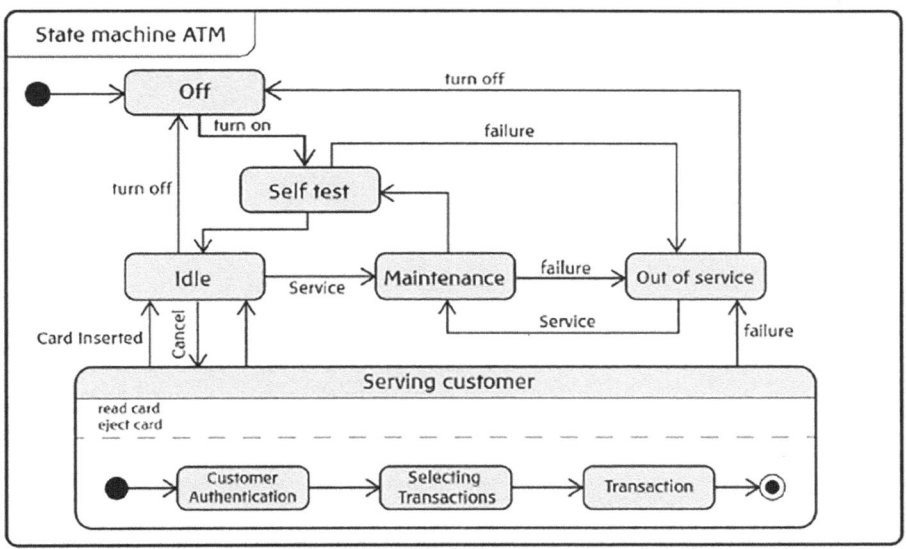

Figure 18-5. *An example of a state machine diagram*

The way we interact with software and how it affects our daily lives is a fascinating subject (at least for some of us!). The next chapter introduces more metaphors to illustrate a few more technical terms. These include user authorization, the question of data storage, and what happens when the software memory leaks.

CHAPTER 19

Metaphors for Technical Terms

Who Are You? What Are You Allowed to See?

A passenger about to board a plane for an international flight usually needs a passport and a ticket. The passport identifies who you are, and the ticket confirms which plane you're allowed to board. Two common terms in accessing a software application are *authentication* and *authorization*. The passport would represent authentication: Who are you? And the flight ticket would represent authorization: Which plane are you allowed to access?

In more formal terms

- Authentication is the process of determining the identity of a user using a service.

- Authorization is the process of determining which rights and access are available to a user who has been authenticated.

© Jonathan Peter Crosby 2023
J. P. Crosby, *The Business Manager's Guide to Software Projects*,
https://doi.org/10.1007/978-1-4842-9231-0_19

Winter Stock—Caching Data

Squirrels store nuts and acorns to eat in the winter when resources are scarce. Foxes, in contrast, continue to hunt for fresh food throughout the colder period.

In software projects, *caching*, like the squirrel's storage of food, is a method of storing a copy of data that allows a certain measure of independence and faster access—but the data can quickly become outdated.

Note   Do you need the data on a daily, weekly, or monthly basis? How important is it to have real-time or near-time data in your application?

Alternatively, your software could follow the fox's example of getting food when needed. Retrieving data on an ad hoc basis using a live interface will allow you to collect the required data as needed. This method means your application will be highly dependent on the other system though.

Note   What happens if the source data system is down? Will your system be unavailable too?

To Wait or Not to Wait

In software, one important factor affecting most parts of any software application is timing, and this involves waiting. Some processes are very quick, while others can be long-running. It's important to make the correct choice on how one piece of software code passes on an action request to another part of code—it depends on the expected duration of the task. Generally, the terms *synchronous, asynchronous,* and *fire and forget* are used to describe these options, which are explained in the following.

Boomerang—Synchronous

When you throw a boomerang, you should stay put and wait to catch it. A similar process takes place in software when the program asks for data and then waits until the data is retrieved, which means it is done in a *synchronous* manner.

Doing Something Else in the Meantime—Asynchronous

An example of an *asynchronous* process is playing chess against another person online. Once I've made my move, I can go off and do other things. I'll get a message once my opponent has made their move. In software, the code sends a message, and instead of waiting, other tasks can be done in the meantime. This technique is used to allow more users to access the same computer system simultaneously. Both in synchronous and asynchronous cases, the requestor expects a response.

Fire and Forget

This term comes from firing missiles and refers to a stage where a missile no longer requires any further guidance during launch. This technique eliminates waiting time as it allows the next missile to be fired before the first one hits its target. Similarly, in software, *fire and forget* means that the originator doesn't expect a response. However, the originator will not receive direct information if anything goes wrong. Nonetheless, this method can be useful for background jobs such as sending out registration emails.

Airport Conveyor Belt—Memory Leak

At the airport, a rotating conveyor belt is overloaded with luggage (Figure 19-1). More and more suitcases are being piled onto it, while too few people are there to claim their luggage and free up space. This is a simplified analogy of a common software issue called a *memory leak*. These leaks on a computer system come when a program doesn't manage the memory correctly, neglecting to remove data that is no longer needed for instance.

Figure 19-1. *Freeing up space*

Switch Off the Lights—Bits and Bytes

A *bit* is a kind of an on/off switch where "0" means "off " and "1" represents "on." Inside a computer, fine cables laid in parallel send "on/off" signals simultaneously. When 32 cables are placed next to each other, this is called 32-bit, 64 parallel cables are therefore 64-bit, and so forth.

A *byte* is a group of 8 bits. Groupings of bytes on a computer are formed into commands that are understood by the computer but are hard for humans to read; hence, most of the code written by developers is done in a so-called *third-generation language* (3GL). These programming languages use natural language instructions that humans find easier to understand and use to write most code. Once the development is completed, the code is converted into a language that computers can read.

In the following chapter, I'll briefly mention some of the areas that need extra care to get right—it's useful to know about them in advance. These areas include the complications that can arise from the differences in dates and times in different places, for example, and the challenge of providing an effective *search* feature on a website.

CHAPTER 20

Tricky Areas in Technical Development

It can be useful to check through the following list of technical elements to identify whether your project will include any of these slightly tricky areas. The list is by no means exhaustive, but these are the most common ones I've encountered. Ensure that your planning caters for these elements and that you allocate sufficient time for tackling them.

Dates

Times and dates can be very complicated, especially if the software is used in different time zones. Dates are already confusing due to the different display formats. The date 06/08/2017, for instance, could mean June 8th in one country and August 6th in another. Some countries even have a completely different calendar. Daylight Saving Time (with clock changes) applies only to some parts of the world, and many countries change the time on different dates. Additionally, official holidays vary across countries and regions too. Many of the support issues I've been involved in have been related to times and dates.

© Jonathan Peter Crosby 2023
J. P. Crosby, *The Business Manager's Guide to Software Projects*,
https://doi.org/10.1007/978-1-4842-9231-0_20

Note   Will your system calculate the age correctly for someone born on the 29th of February in a leap year?

State

The state of a system is hard to manage, especially in a large system with many active users and processes. Ensuring that state is managed well, even after a system crashes, is a challenge.

Cache

Getting caching right is one of the biggest challenges in software. Caching is used to make things faster, which is generally a good thing. The hard part is in deciding how long the data is valid for. When viewing live data from the stock exchange, for example, the user doesn't want to see old stock prices.

Spot the Difference!

Let's say the old finance system is being replaced by a new one. Some of the reports originally produced in the old system need to be identical in the new system. Is the monthly report for every customer still correct? Why are there differences? Often when systems are replaced, the figures don't match. The analysis to find out which of the two systems has a bug can be tedious.

Interfaces to Legacy Systems

Being dependent on a supertanker system can require a lot of foresight and cause additional costs as well. Modern systems are quick at setting up new interfaces or amending existing ones. Conversely, a legacy system may have very few releases each year and could be expensive to adapt.

Testing and Test Data

Creating meaningful manual and automated tests can involve a great deal of work. Systems with a lot of complex business logic can necessitate an abundance of special test cases. What's more, ensuring the test data is of high quality requires a considerable amount of time. The test data should be similar to production data, both in form and quantity.

Messy and Undocumented Code Base

Having to clean up, rename, and figure out many lines of code will mean a substantial amount of work for the development team. Using concise language in code from the start helps avoid this. Future code changes should also involve fixing technical debt and refactoring the code to better reflect the business model.

Search Feature

Returning the search results the user wants to see is incredibly hard. Many websites offer their own search feature, but not many do this well.

Performance

Software performance is a wide topic—there can be so many reasons for poor performance. Keep in mind that the speed of a website or application can be a subjective matter. Therefore, it's best to discuss and agree upon the expected speeds as part of the non-functional requirements. In most cases, performance improvement is a gradual process because it involves a series of small adjustments rather than one major issue.

This wraps up Part 3, "Technical Guide," on the technical topics. Well done if you've worked through it all—I hope you've found some good takeaways! If you're still hungry for more (food pun intended), you'll find some further interesting topics to devour in the appendixes after the brief summing up that follows next.

CHAPTER 21

To Sum Up

This book has taken you on a journey where first, with the help of metaphors and analogies, you viewed the core concepts of software projects. We followed that with a look at the more practical side of projects and dug a little deeper into the technical topics. Armed with this newly acquired knowledge on the various facets of building software solutions, you can now walk into your next project meeting with confidence.

Familiarity with some of the most common technical terms will also help you make a positive impact on your next or even on your current project. A sound understanding of the most common terms will, in fact, greatly assist you in communication and decision-making.

If I had listened to my uncle, a retired software engineer, I might never have written this book. He tried to dissuade me, saying jokingly, "A little knowledge is a dangerous thing!" Similarly, as Joel Spolsky explains, "I use the term 'Econ 101' a little bit tongue-in-cheek. For my non-American readers: most US college departments have a course numbered '101' which is the basic introductory course for any field. Econ 101 management is the style used by people who know just enough economic theory to be dangerous" [Spolsky 2006]. Now that you know enough about software projects to be dangerous, I truly hope you'll use this knowledge wisely. Joking aside, I believe it's very important to empower everyone involved in a project to contribute effectively. Having the business and management not only on board but actively involved in contributing to good team decisions is a huge benefit. Not everyone in IT will have the same level of

© Jonathan Peter Crosby 2023
J. P. Crosby, *The Business Manager's Guide to Software Projects*,
https://doi.org/10.1007/978-1-4842-9231-0_21

formal education. Recently, I spoke to a software developer with 25 years' experience who had never heard the term *technical debt*—by reading this book, *you* now know about it. In some cases, by bringing a fresh perspective, you may even be able to identify deliverables or process steps that an IT professional has omitted. If you feel something is particularly important to the project, you may need to insist on your point or get a neutral opinion.

It's very hard to get everything right in software projects, but I feel there's a huge need to bring the knowledge of how things should ideally be done to a wider audience. At the same time, the best practices aren't written in stone and will often change over time. I've worked on all kinds of projects, and some have come very close to being perfect. They were a lot of fun to work on, and it does make a huge difference if most of the stakeholders are in the know.

Projects should use an inclusive approach, allowing everyone to freely express their opinion. Democratic decisions work particularly well if the whole group understands the consequences of the decisions being made.

Two areas particularly stand out for me. First, keeping up a high level of communication throughout the project is of the utmost importance. Second, having the understanding and support of upper management in software projects is becoming ever more critical to the success of the entire business.

I hope the knowledge you've gained from this book will equip you for more involvement and fun in your future software projects. The goal was to help you grasp certain aspects on the subject much more completely. I wish you the best of luck and encourage you to engage yourself wholeheartedly in your next project.

APPENDIX A

Collaboration

Roles in Software Projects

The roles required in a project will vary depending on the size and formality of the project. In smaller projects, one person may take on multiple responsibilities, whereas in large ones, roles can be even more granular than the ones listed in the following. Most projects assemble a core team and rely on the help of an extended group of experts.

First, we'll look at the general roles and then at the Scrum methodology roles. Alternative role names may be used depending on your company or industry. I've listed only the main duties here, although even these can vary slightly from project to project.

General Project Roles

The **business analyst** focuses on the needs of the customer, managing the requirements, and creating test cases for the new features.

The **enterprise architect** ensures the company's overarching system architecture rules are upheld.

The **designer** creates the user interface (UI) mockups, UI designs, and possibly even user experience (UX) designs.

The **developer** produces the low-level design document (LLD), writes code, and generates code tests.

© Jonathan Peter Crosby 2023
J. P. Crosby, *The Business Manager's Guide to Software Projects*,
https://doi.org/10.1007/978-1-4842-9231-0

The **dev lead** manages the development team, coordinates their work, and helps resolve any impediments.

The **product manager** is responsible for the road map and the new features of a software solution.

The **project manager** leads the entire project, coordinates with all the stakeholders, and is ultimately responsible for the project outcome.

The **solution architect** creates the high-level design (HLD) of the system and ensures these guidelines are adhered to.

The **sponsor** is responsible for the financial aspects and scope monitoring as well as for steering the project out of difficult situations.

The **subject matter expert** contributes expertise on specific business matters and helps shape the software's business logic.

The **test manager** tests the software, manages the test cases, and coordinates all the different test persons.

The **tester** performs the tests on the software according to the defined test cases.

The **user/user representative** helps by defining the user requirements and performing the user acceptance testing.

Scrum Project Roles

The **development team** is self-organized, makes decisions as a group, and works on tasks as described for the preceding developer.

The **product owner** coordinates between the team and the stakeholders, gathers software requirements, and is responsible for the outcome of the project.

The **Scrum master** coaches the team, ensures the Scrum rules are upheld, and tackles impediments.

Manifesto for Agile Software Development

More than 22 years ago, a group of 17 experienced software developers met with the goal of improving software project success. They all had many years of hands-on experience, and most of them were already published authors. Their meeting in Snowbird, Utah, had a huge effect on how teams work together.

> *The manifesto was published in 2001 by a group of experienced software engineers with the goal to increase software project success and be more adaptable to changing requirements:*
>
> *We are uncovering better ways of developing software by doing it and helping others do it. Through this work we have come to value:*
>
> > **Individuals and interactions** *over processes and tools*
> >
> > **Working software** *over comprehensive documentation*
> >
> > **Customer collaboration** *over contract negotiation*
> >
> > **Responding to change** *over following a plan*
>
> *That is, while there is value in the items on the right, we value the items on the left more.*

Kent Beck	*Ron Jeffries*
Mike Beedle	*Jon Kern*
Arie van Bennekum	*Brian Marick*
Alistair Cockburn	*Robert C. Martin*
Ward Cunningham	*Steve Mellor*

Martin Fowler	*Ken Schwaber*
James Grenning	*Jeff Sutherland*
Jim Highsmith	*Dave Thomas*
Andrew Hunt	

Principles behind the Agile Manifesto

Our highest priority is to satisfy the customer through early and continuous delivery of valuable software.

Welcome changing requirements, even late in development. Agile processes harness change for the customer's competitive advantage.

Deliver working software frequently, from a couple of weeks to a couple of months, with a preference to the shorter timescale.

Business people and developers must work together daily throughout the project.

Build projects around motivated individuals. Give them the environment and support they need and trust them to get the job done.

The most efficient and effective method of conveying information to and within a development team is face-to-face conversation.

Working software is the primary measure of progress.

Agile processes promote sustainable development. The sponsors, developers, and users should be able to maintain a constant pace indefinitely.

Continuous attention to technical excellence and good design enhances agility.

Simplicity—the art of maximizing the amount of work not done—is essential.

The best architectures, requirements, and designs emerge from self-organizing teams.

At regular intervals, the team reflects on how to become more effective, then tunes and adjusts its behavior accordingly.

—[Manifesto for Agile Software Development 2001]

More on Communication

Activating Our Senses

On the topic of understanding, one interesting piece of research by Tor Nørretranders reveals the processing bandwidth of the human senses when mapped to common technical devices. Sight is our strongest sense by far. This fact emphasizes the value of visual aids in creating a common understanding on data and systems. Unsurprisingly, perhaps, we can process far more information using our sight than with our other senses.

SIGHT

TOUCH

HEARING / SMELL

1250 MB/s
Same bandwidth as a computer network

125 MB/s
USB key

12.5 MB/s
Hard disk

TASTE

Bandwidth of our senses by Tor Nørretranders

Choosing the Right Communication Channel

You'll need to find the most effective ways of communicating in your project. It's worth discussing this matter with your teams to establish the best methods for them. We have plenty of choices these days, such as chat tools, emails, meetings, phone calls, and bug tracking and document management systems. Establishing good communication channels will help you keep track of all aspects of the project. Otherwise, finding out which data and which decision is where can be a daunting task. Defining some soft rules will minimize the disturbance to others at work and help avoid unnecessary meetings, as well as reducing the number of emails sent and so on.

Talking to someone who's focusing on a difficult task or using the phone within their earshot can be very disruptive. Such interruptions can make the person lose their train of thought completely. To minimize disturbances, avoid asking separate questions at different times—note them all down and discuss them in one go with the people concerned. Also avoid sending long email chains addressed to dozens of people. This practice is usually a suboptimal form of communication. Further, disturbing a meeting between two or more people just to ask a question that isn't urgent is something everyone should avoid. Everyone, including managers, should be obliged to join the queue, unless it's a really pressing matter.

Also think about the form of communication that's best suited to the size of the task. Sending a two-sentence email is fine for a 5-minute job. Too often, though, I've received such requests for work that required several days to complete. In such cases, and out of respect for the other person's time, it's best to communicate verbally, sitting with the other person if possible to speak about the task. This face-to-face discussion will be a good opportunity for answering any open questions and will increase the chance of the work being done right the first time around.

One very effective way of sharing information in a large project is through a science fair. At one such fair I attended, four small groups presented their specialized topics from a project to the other project members and stakeholders. The audience watched each 10-minute presentation before taking part in a 5-minute question-and-answer session for each one. I found this a fantastic way to gather a lot of information in a very short time.

During one project, we organized an event entitled "A pinch of subject matter knowledge for our new colleagues." Everyone prepared and then presented information on one sheet of paper to describe a given term using the following points:

> *Definition*: What is it?

> *Characteristics*: Values

Implications

What other information helps describe the term?

Also, when new colleagues join the team, don't forget to introduce them to the other members. This simple and effective gesture is sometimes neglected—maybe the new person started during your vacation, or you were in a meeting when they introduced themselves to the others. Depending on the size of the project, it can take quite some time to speak to everyone. Nothing beats personal introductions though.

Note   Has everyone on your project been properly introduced? Are you familiar with the various responsibilities of each team member? Have you spoken to the key contacts of other systems you'll be communicating with? Have you ever sat next to people at work whose names or roles you didn't know?

When communicating with another team, it's often best to contact the team rather than an individual. Using a specific team mailbox allows all team members to share the workload and knowledge more easily.

Meetings

Golden rule: Arrive well-prepared at all meetings.

I've never heard anyone say, "I love meetings," but in most workplaces, a certain number are necessary. A bit more planning would make them a lot more tolerable. Sending a brief meeting agenda in the invite makes a good start. At the beginning of a meeting, as one of my early bosses taught me, briefly state the goal of the meeting. Remind people of this goal if things move off track. If colleagues are getting bogged down with too much detail at any point, try asking questions such as "Can we take a step back here?" or "What problem are we really trying to solve?" These kinds of interjections serve to redirect the focus to the matter at hand.

Incidentally, when sending meeting invitations, it's best to send important project decisions or information in a separate email. If anyone who cannot attend the meeting declines the invitation, most current email tools result in the contents of the invite being lost.

Presentations

Well-prepared presentations that are engaging and are customized for the audience are something I really enjoy. Usually much better organized than meetings, presentations can convey a lot of relevant information in a short time. A good presentation can be inspirational. It's essential that the presenter speaks loudly enough for everyone to hear. Also, during question time the presenter should repeat every question asked by an audience member so that (1) the presenter can confirm the question has been understood correctly, (2) they gain a little more time to gather their thoughts, and (3) the people furthest away are sure to hear the question.

Don't overload the slides at presentations with too much text or too many visual elements—it simply suggests you don't know how to prepare a presentation. This bad practice is so common it hurts. Yes, the people at the back should be able to read it too, and apologizing doesn't put things right. Split things up across slides, magnify them, or use another format. If an item is just too big, don't try to squeeze it into a slide, but use a placeholder, and present the topic verbally instead. As a rule of thumb, a slide shouldn't contain more than seven concise bullet points. Some people have their own reasons for overloading slides, however, as I learned from a short dialogue with a project manager:

> **Project manager**: Can you please send me that system overview for my presentation?

Me: Sure, but no one in the room can read what's on the projector screen.

Project manager: Exactly, it looks nice and complicated. Just to show them how busy we are.

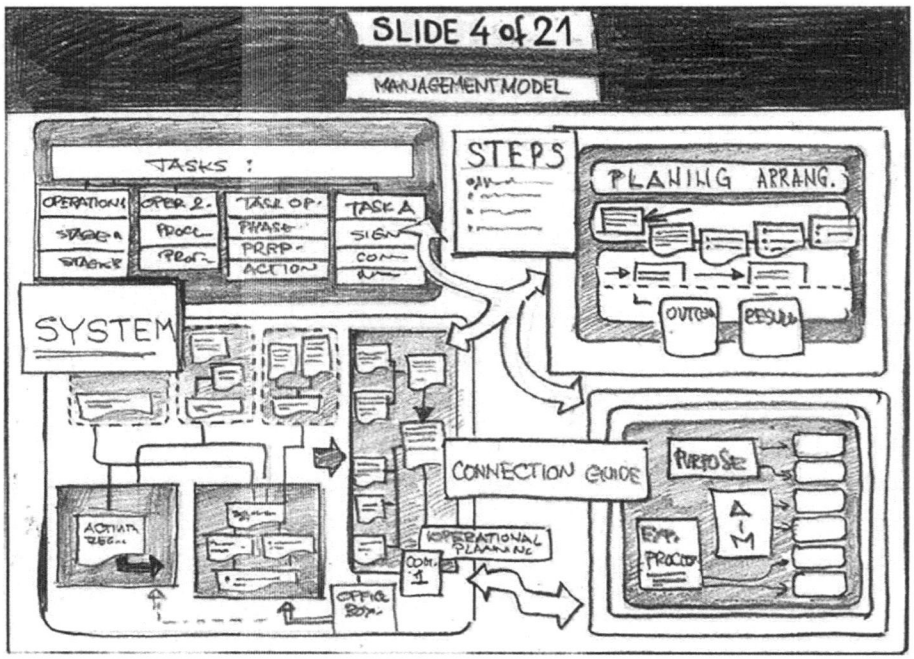

Overloaded slides are impossible to read and don't impress anyone.

High-Speed Talking

A foolish way of trying to convey an impression of competence is to speak very quickly, either with or without using jargon. Some people appear to fall for this one. From my own observations, there seems to be no correlation between fast talkers and the quality of their message. The audience will have trouble following a talk that's delivered at top speed—they're liable to switch off as the barrage of ideas rushes over their heads. Also, it's very hard to ask a question when there are no gaps between sentences. Having good speaking skills is a big asset. It means presenting ideas effectively and *pausing* from time to time to check that everyone is following. An effective speaker allows everyone present to absorb the new information, giving them time to build up a picture on the topics presented. They make the audience feel at ease and ready to engage in a dialogue. Also, just as having the quieter, more thoughtful type of people on the team is very important; giving them the chance to speak too is critical to the project outcome.

Powerful Voices

Even though this factor is rarely mentioned, people with strong persuasive voices undoubtedly have a huge influence on the listeners and on the decision-making process. But just as with high-speed talking, a voice that captures your attention may not always indicate quality in the actual content of the talk. Here lies a danger. A speaker may use this asset to serve their own interest rather than that of the team or company. It's therefore important for the audience to be aware of this possible influence and always take care to focus on content over delivery.

You've probably encountered the one who often tries to monopolize meetings. Understandably, team members can have difficulty in determining whether this character really does have the sharpest insight, as they seem to claim. Recent research indicates that the loudmouths often lack expertise, in fact. This study from the University of Utah and Idaho State University also shows which groups perform best in problem-solving tasks—it turns out that the ones taking more time to identify the actual experts are the ones that come out on top. "We'd hope that facts would be the currency of influence," says Bryan L. Bonner, main author of the study and an associate professor at the University of Utah. "But often, we guess at who's the expert—and we're wrong."

Prof. Bonner describes how people tend to look for the wrong signs—such as a speaker's confidence level, gender, and/or race—when making judgments about expertise. They do this instead of focusing on the actual content of the speaker's contribution. These findings apply mainly to group problem-solving tasks such as estimating the number of product units a firm will ship the following week. Other examples include valuating a company as well as less concrete tasks like generating ideas for a marketing campaign [Bonner & Bolinger 2013].

A good sense of humor is another powerful asset. This trait can work wonders when it's used in a positive manner to unite the team. It's great to have fun at work and at team events outside of work too. I've also seen how this talent can be misused though. The danger again comes when the speaker uses their influence on the group for their own personal ends, although they may do this without even being fully aware of it. Using their good relations with certain team members, the speaker can divide the group by undermining others they dislike.

In other words, a speaker's ability to captivate an audience can be used either to unite or divide the team or to prevent other team members from speaking up. The best group dynamics come when everyone feels comfortable enough to express their opinion and contribute constructively to a project.

Knowledge Sharing

Avoid building up islands of knowledge that are inaccessible to everyone else. I've seen some very smart IT guys trying to make themselves irreplaceable by keeping their knowledge to themselves—but managers soon become aware of this. It doesn't make for a fun or a smooth-running working environment when some people keep their knowledge to themselves.

Many companies benefit from their employees' great wealth of experience, and executives usually value experience as an important resource. However, it is a resource that is rarely utilized fully or promoted systematically. This tendency is shown in a study conducted by the Rheinische Fachhochschule Cologne. The results show that many of the 600 executives surveyed knew the value of knowledge combined with practical experience. Not all their companies ensured this was fully utilized or shared, however [Bruns, Eschenbach & Maier 2016]. Try to make sure your project benefits from the valuable input of the experienced persons in your company. Encourage everyone in your project to share their knowledge.

Givers and Takers

Understanding people's characters can be very helpful for managers and team leaders. Managers can now take advantage of recent research that reveals the dynamics within teams. Acquiring a basic understanding of the different character types can greatly help them make the most of their team members. As Adam Grant points out in his TED Talk on team dynamics, you'll find three basic types in any group of people. Each group responds in a different way to the others. First, there are the ones who give and the others who take. By observing your team members well, you can identify and support the *givers* and root out the *takers* to build a team that is more effective and efficient. Further, this research identifies a third category of people—the *matchers*. These individuals, who make up the majority of people, try to strike a balance between giving and taking, often viewing it as an exchange of help or favors. They find it hard to tolerate the takers who don't play the game.

The givers spend time helping others at the expense of their own work and hence score badly in terms of productivity. Yet organizations with high proportions of givers excel in customer satisfaction, as well as having higher profits and even lower operating expenses, to mention only a few of the benefits.

The successful giver is also willing to receive help when necessary. Matchers will become givers when others ask them for help, and giving is most often done in response to a request. It turns out that the majority of people actually enjoy being helpful and will respond very well once asked. Allowing a culture of successful givers to flourish does, as already mentioned, mean removing any takers from the team. The takers will only create resentment and will discourage the matchers from giving. As these takers are liable to hide behind an agreeable surface manner, you need to observe and question them well to discover their real motives [Grant 2016].

Business Teams Making Their Own Tools

Some of the software projects I've worked on over the last two decades were originally built using Office tools. These projects may start off like this because no budget is available for a more sophisticated approach. These systems subsequently grow organically over the years until finally being replaced by a full-fledged system at some point. Such solutions often originate from a business department.

I believe these quick and pragmatic solutions do give business teams some independence. Such solutions can even grow to become business-critical. In some cases, however, the business teams may not wish to hand over the responsibility to the IT teams, viewing them as slow, expensive, and lacking the necessary business knowledge.

Let's look at why there is such a discrepancy in time and costs between an Office solution being built by an end user and a solution being built by the IT department. Coordinating with the IT team will take up valuable time for one thing. Additionally, the IT department will need to consider aspects of the solution that a business department may not have paid much attention to. These include technical documentation, data backups, security concerns, user access management, disaster recovery, and various other standard best practices. In my experience, many Office applications created by users lack some if not all of these criteria. Not all applications run by IT departments are well-managed or documented either though, but overall, these do tend to be handled better than Office solutions. Proceeding in a structured and compliant fashion ensures that we build software on a solid foundation.

Should your business-critical Office applications be analyzed and inventoried by the central IT department? Are some of these applications candidates for being converted into professionally developed solutions? Do the IT departments need to gain a better understanding and show more appreciation for such Office solutions created by business teams?

For me, the answer to the final question should be a loud "Yes!" These internally created applications are a rich source of information that can help the IT department understand most user requirements implicitly. Somewhat mistakenly, many formally educated IT professionals will frown upon such applications for failing to meet their high standards. Yet understanding where these solutions come from and the opportunities that can emerge from them is actually very useful.

In one case, the team leader in the risk department created an Office solution and asked their team members to use and extend it too. Some team members expressed their wish to transfer the solution to the IT department, especially as the solution was largely undocumented. By creating large and undocumented Office solutions, the risk department was ironically exposing the organization to further risk.

It's quite natural that the business department may fear that by handing over the ownership of their data, the IT department will somehow "hold them hostage." IT departments therefore need to be more open to pragmatic solutions and embrace the business initiatives. These tools will continue to

263

be created by business teams. For mid- to large-sized companies, I'd even recommend creating teams that specialize in supporting and sometimes taking over responsibility for these Office solutions.

Hopefully, this section has given you some insight into both sides of this topic that affects nearly every company around the world. We can't ignore the way companies and teams work. Rather, we should embrace the status quo and identify anything that requires a more stable basis.

APPENDIX B

Glossary

The technical terms used in the book are described here in plain English. Although these terms may have different meanings depending upon the context, the terms here are defined in the context of software projects and IT in general.

An **administrator** is someone who has leveraged rights to a software application, who can typically perform user administration and has access to detailed information about the application to provide support.

The **Agile methodology** is a rather informal approach to teamwork with the goal of delivering software with high business value and fully functional features in shorter timeframes. This methodology can be used in other types of projects too. The teams are self-organized and flexible and communicate closely, both within the team and with all the stakeholders.

App and application both describe a finished piece of software running on a device. An app will typically perform one main function, whereas an application can contain a range of functions. The line dividing software apps from applications has always been a little blurred, and this is even more the case now that smartphone devices and apps are becoming ever more powerful. A software application is also referred to as a program. These terms all loosely describe similar things.

The **application lifecycle** encompasses the entire process of producing software from defining requirements and new features, designing those features, and then coding and testing them. Finally, these enhancements are made available to the users by going live, and maintenance begins. A new lifecycle will start afresh with further requirements and features.

© Jonathan Peter Crosby 2023
J. P. Crosby, *The Business Manager's Guide to Software Projects*,
https://doi.org/10.1007/978-1-4842-9231-0

An **asynchronous call/request** occurs when a computer or application requests data from another system. More specifically, *asynchronous* describes the way in which the data is requested. The application or computer can continue doing other things until it is notified that the data has arrived (see also **synchronous call/request**).

The **audit log** stores the details of financial transactions in chronological order in a log file (see **logging**).

Authentication is the process of identifying the user who is requesting access and granting as well as denying their access.

Authorization ensures that the user is allowed to access a certain resource.

A **backup** refers to copying and storing data in another location to prevent data loss.

The **back end** is the term used to describe the server side of an application. This is the part that is not visible to the user and typically contains the business logic, data storage, email distribution, and various other services (see also **front end**).

The **bandwidth** refers to the maximum speed of a network connection. The speed is affected by all the devices that are connected in the chain.

Bit/byte: A bit is a binary digit that goes on and off like a light switch. A byte is a group of 8 such bits. All digital devices and storages are a massive collection of bits.

A **black box** is a system where the internal workings are highly obscure or non-transparent. These are very hard to support as it is unclear how and why actions occur. Such black boxes should be avoided.

A **breaking change** refers to the part of a new software version that is incompatible with the previous version.

A **browser** is the software used to look at websites. Examples are Firefox, Safari, Chrome, Internet Explorer, Opera, and so on.

A **bug** is the term used to indicate a software failure or fault when the app, program, or system is not working as expected.

Business logic is the part of a software application containing the rules specific to that business. The logic could, for instance, be based on the terms agreed in a contract between a company and a customer, such as the conditions of a wireless phone plan between a carrier and a customer.

Business requirements are the underlying needs of the business to fulfill its tasks. The requirements should describe the business goals to be achieved.

Caching means to store data locally or in another place where the data can be retrieved more quickly. The difficulty in caching is that the stored data can quickly become out of date. A stock price can change very quickly, for instance, and caching that data would mean the user may see an old price. On the other hand, something like a company logo image on a website doesn't change very often. A logo would be a good candidate for caching as the image wouldn't need to be sent across the network for every page view.

The **client** is the application or system requesting the data or service from a server. When google.com is typed in, for example, the client could be a laptop, PC, tablet, or phone. On the other side, google.com would be the server answering your request (see also **server**).

Code includes commands and instructions to form a piece of software.

Code check-in/commit: A programmer frequently performs this task during the software development phase. This process involves adding the changes to the code already made on their device to a shared code location for other programmers to use.

In a **code review**, a programmer shows the code changes they've made to another programmer for their approval. To allow for changes to be easily traced, the name of the reviewer, the date, a brief description of the change, and a link to the requirement are all stored.

Coding is the common term used to describe the process of writing software code.

The term **compile** refers to the process of converting code from a human-readable form to a computer-readable form.

The **compiler** is a piece of software that does the translation from human-readable to computer-readable software code.

Components are the various parts that make up a piece of software. Each component will typically have a main responsibility and will itself be a grouping of multiple features. The user interface and database are examples of components.

Continuous deployment is the task of adding the changes to the live/production system(s) shortly after the code has been written and successfully tested. Most if not all tests are done automatically, and new changes could be added daily, for instance.

Continuous delivery describes the setup when the business departments decide on the times to add new features to the live/production system(s). Continuous delivery relies on continuous deployment to a production-like environment, and the business simply activates the new features.

Continuous integration refers to a programmer frequently adding small code changes to the shared code location for other programmers to use. It can mean that code is integrated multiple times every day. This method prevents a team of programmers from each making large changes that might not be compatible.

Cookies are small text files stored in the browser containing information about the user's visits to a website.

CRM stands for customer relationship management. This system is used to store customer data and customer communications.

Data anonymization is a process that involves masking the confidential data when copying it from a live system to a test system, for instance. The testing team may not be authorized to see the real customer names and addresses, so this data is modified to ensure confidentiality.

A **data center** is a physical location where servers that store data are stationed. These server rooms are often specifically designed for this purpose and are cooled to the optimal temperature for running these computers.

Data flow is the term used to describe the route and direction data takes within a system or across multiple systems. The data flow is often represented graphically to aid understanding.

Data migration describes the data that needs to be transferred from an old system to a new one. The data may also require transforming so that it matches the shape of the new location.

A **database (DB)** is typically a central location for storing data in a predefined structure. Many systems use some form of database to store data.

Debugging: The programmer does this continuously while coding to identify mistakes/bugs in the code.

The **default value** is the value prefilled in advance that can typically be changed by the user. When you use a holiday website, for example, the default value for the travel date might be set to two days from today.

Definition of awesome is used to define the goal or the solution to a problem in the context of an ideal situation without limits or constraints. It helps the team and the stakeholders think outside the box.

Definition of done: The team decides collectively when work is considered as done and gets marked as such. This may sound like a simple and obvious thing to do, but it can involve disagreements and lengthy discussions. There are many possible criteria involved in this decision, such as whether the task is completed once it compiles correctly on the shared development server, once it has passed all the automated tests, when the business confirms the acceptance criteria are met, and so forth.

A **deliverable** refers to a distinct part of the software that is seen as an individual item created for a user, such as a report, a new screen, a document, or in some cases the whole application.

Deployment refers to the installation of the changed software on the server(s) or client(s).

The word **deterministic** describes a program that will always give the same result when given the same input. If you had a piece of code that added two numbers, for instance, the result would always be the same when the same two numbers were entered.

A **developer** is the person who writes software code.

Development is the process of writing code.

The word **DevOps** comes from *development* and *operations* and is used to describe the overlap of these two areas of work, where the respective teams collaborate closely. The development team is responsible for writing the code, whereas the operations team installs the software on the server(s) as well as taking care of the setup and availability required for the technical infrastructure.

The concept of **disaster recovery** refers to how a system is restored to be fully functional again after a major incident has occurred. There are many possible causes of a disaster that could affect one or more systems. Many companies have detailed documentation of how to handle such an incident. In the case of many systems being affected by a disaster, the order of restoration will be very important. The network is usually one of the first elements to be restored, followed by other central infrastructure and mission-critical systems, depending upon their importance.

A **domain** refers to the area of business in which the project is situated.

A **domain model** is the visual representation of the real-world business elements and their relationships. This model is used to enhance communication between the users and developers. Having a clear domain model ensures everyone is referring to the same things.

DRY is an acronym that stands for *don't repeat yourself!* The term is used in programming to avoid repeating any of the code. This is very useful when something needs to be changed, as the code only needs adapting in one place.

The **end user** is the real person using the software.

The **environment** describes a group of connected systems that together provide all the functionality of the application. The live/

production environment may consist of a single database and web server or of numerous such servers. The test environment usually has its own set of servers that are set up identically to servers in the live environment. There may be additional environments, such as development, education, and more, depending on the size and criticality of the application. These other environments are not usually set up identically to the live environment for cost and maintenance reasons.

Fail fast is an approach that can be used in code to make any errors visible and allow them to be addressed sooner rather than later, thereby making the code more robust.

A **feature** is any part of the software application that can be considered and discussed individually, such as the search or payment feature.

The **front end** is the area in software the user will interact with. It is usually the visual part of the desktop application, phone app, website, or other user interfaces (see also **back end**).

The **functional requirements** reflect the desired features of a system that achieve the expected behavior from a user perspective. In other words, the functional requirements should answer the question of what a system should do (see also **non-functional requirements**).

The **functionality** describes the tasks that a piece of software can or should fulfill.

Greenfield project is used both as a term and a metaphor to describe a project creating brand-new software, as opposed to extending or improving existing software.

A **heat map** is a graphical representation that shows the parts of an application or website where the users have clicked or pressed. The more clicks that are made on a given area, the brighter or warmer the color will be, as displayed on this analysis tool.

High-level design (HLD) is a document that contains an overview of the system and includes the blueprint of the main components translated from the requirements. The software will be developed from this blueprint (see also **low-level design**).

271

Identifiers/unique identifiers are used to represent, store, and retrieve a single unit of data. For instance, an individual order on a shop website could have a specific ID number that may or may not be visible to the customer. The ID may be used only internally within one system or may be shared across multiple systems. In most cases, this ID would never change.

Impact analysis is a task performed to identify the persons and systems affected when a change is planned. The impact analysis should be done early in a project to identify all stakeholders. Additionally, an eye should be kept on new changes during the project that affect new stakeholders.

The term **impediment** is used to describe an obstacle blocking the progress on a task.

The **implementation** is the actual realization of planned work.

Incremental refers to the gradual adding of changes to a system.

Infrastructure is a term that generally means the underlying hardware and software required to run the software being developed by the project. The server required to run a website or application would, for instance, be referred to as part of the infrastructure.

An **interface** is a broad term that describes the intersection between two or more separate items. There could be an interface between two pieces of code, for example, between two systems, between the user and a system, and so on. The interface can also be seen as the contract between the separate parts that defines the predefined structure of communication.

The **Internet of Things (IoT)** refers to devices that may not have previously been connected to the Internet, such as washing machines, microwave ovens, light switches, and weather sensors.

Information technology (IT) is a general term that refers to computers, to the processing and storage of data, and to software, the Internet, and various other related areas.

Iteration and iterative are the terms used to describe any process that needs to be repeated to reach the desired goals. For example, the so-called *sprint* task in Scrum methodology is repeated on a regular basis, perhaps weekly or biweekly.

Kanban is a management method that can be used both in software projects and in other areas. The method focuses primarily on leveraging work in progress (WIP) limits (see also **work in progress**).

The **Kano model** offers a structured method for identifying customer satisfaction.

A **legacy system** is an older piece of software that may have originally been state of the art, but no longer meets that standard. Although the software is now outdated, it may still play an important role in the company. A new system may need to receive data from the old system. If the new system will replace the legacy system, all relevant data must be transferred.

Low-level design (LLD) refers to the document containing a granular description of the software components being built. The descriptions may show the detailed structure of data and an outline of the code (see also **high-level design**).

Load balancing is the method used to evenly distribute concurrent connections. The systems typically fulfill the identical functionality for all connections. Two or more systems will share these connections made by users or other systems.

Logging involves the process of writing detailed information on system usage in log files while an application is running. The data collected assists the support teams and enables monitoring systems to collect this information.

Macros are commonly known in Office applications for automating repetitive tasks or groups of tasks in a manner that is simpler than writing code.

The **Manifesto for Agile Software Development** was defined by a group of highly regarded software professionals in 2001. The Manifesto combines many existing best practices to help improve software projects.

The goals include better communication, collaboration, functioning software, and handling changes.

A **master system** is the one that has the core responsibility over a certain group of data or services. A company that has various systems containing address data may find it best to use one as the master system that delivers the data to the other systems.

A **memory leak** can occur when the information inside a computer takes up an ever-increasing amount of memory space. The system may run out of free memory space, and this may destabilize the system or cause it to crash.

Metadata refers to certain additional information that describes data. The metadata of an image file would contain descriptive information like the file size, the date the image was created, the resolution, and many other details, but not the image itself.

The **methodology** describes the structure of the project and the organization of the people working on it.

Microservices refer to a software architecture paradigm, in which each fine-grained service is responsible for a unique and small business goal (see also **monolith**).

A **mockup** is created to give the users and other stakeholders a detailed and realistic impression of how the final UI could look (see also **wireframes**).

A **module** is part of a software application. This part usually works independently and can even be exchanged if required.

Monitoring involves systematically watching an application. Monitoring software can provide current and historic system information on performance, availability, usage, and errors, among many other things.

A **monolith** or **monolithic system** is a loose term used to describe a large system that is responsible for a variety of different tasks. These tasks are typically interwoven components that are dependent upon each other (see also **microservices**).

274

Minimum viable product (MVP) is the term used in product development to describe the minimum number of features the customers are willing to accept for the product to go live.

Non-functional requirements (NFR) describe the areas like system performance, availability, accessibility, privacy, and other goals related to quality. The non-functional requirements address the way a system acts (see also **functional requirements**).

Off-the-shelf software refers to standard, ready-made software as opposed to the custom or bespoke software developed specifically for a customer.

Open source software can be read, used, and amended by anyone and is usually free of charge. Groups of programmers in different parts of the world will often work on open source software.

The **operating system** is the base software running on a computer and includes products like Windows, macOS, Linux, iOS, Android, and so forth.

Outsourcing is the practice of moving certain internal roles or departments to an external site with the goal of reducing costs.

Pair programming involves two developers sitting together at one desk and working together at coding.

A **pattern** refers to the general approach used for solving a common problem. The development team often decides together on the pattern to be used to solve a specific problem.

Personas are fictional users created to assist project teams in understanding the perspectives of typical users. These personas will usually be given names, identities, hobbies, pets, and so forth to give insights into how real customers think and how they will use the software to reach their objectives.

Pivot tables and charts are aids for visualizing data in spreadsheets and in other specialized tools. The data is usually shown in a grouped and summarized format.

A **power user** is an experienced user. They may not have a formal IT education and don't write software code but have more advanced know-how of a system than an average user.

A **prerequisite** is an essential condition or event that must occur before the desired task can proceed.

Programming is the act of writing software code.

A **programming language** is a set of predefined instructions a software engineer uses to instruct a computer. Most programming languages are written in a human-readable form. For distribution, the code is converted into a computer-readable form.

A **proof of concept (POC)** is a pre-project that is smaller than the actual project and is done before the full project is started to determine whether the underlying idea or theory will work.

Prototype/prototyping is like a PoC. The prototype is created to show how an idea should look and includes elements both of the design and interactivity envisaged for the final product.

A **provider** of software products is an external software development company that either creates custom software or offers standard software products.

Pseudocode is a human language description of what the code will do exactly. This is produced to give team members an idea of what the planned software will do; it cannot be read by computers.

A **query** is used in databases to request data or to perform an action on it, such as updating or deleting some data.

Refactoring is the process of rewriting and restructuring software code with the goal of improving it, typically without changing the actual functionality.

A **release** refers to the new software version made available to the public, to the customers, or to internal employees.

Release notes describe the details of new software being made available. The release notes are also created for an update of existing software containing the information on the most important changes.

Request/response: A request is sent by a person or system to ask for data or for a service. When someone opens the browser and types in a website address such as `www.google.com` or `www.apple.com`, the browser sends the request across the network to receive a response. Both the request and response will contain data. When the websites are running smoothly, the response will contain the content, design, and layout of the website. If something goes wrong, an error message will be sent in the response.

Requirements describe all the tasks that a system needs to fulfill, as well as the expected quality (see **functional requirements** and **non-functional requirements**).

Restoring means reverting to a previous state. Backing up and restoring data could be necessary for disaster recovery, for testing purposes, and for analyzing a support issue (see **backup, disaster recovery**, and **testing**).

A **revision** in a software product is a minor update that should preferably not include any breaking changes (see also **breaking change** and **version**).

Rights refer to the access given to a user or a system.

Rollback is the process of reverting to a previous version of software if the new software doesn't work as expected and the decision is made to go back to the last working version.

Scalability can relate to companies, teams, and the software itself. Will the company or team still perform well if it experiences rapid growth or a higher workload? Software scalability is concerned with how well the software runs with an increase in users accessing and using it.

Scale-out involves adding new servers.

Scale-up means improving the hardware of an existing server.

Scope creep refers to the risk of adding too many new features during a project, causing it to become bloated, overrun costs, and fall behind schedule.

Scrum is an Agile project methodology that helps reduce the complexity of projects and thereby achieve higher business value.

The term **security** refers to the protection of a device or software from being damaged or compromised and guarding against data being stolen.

The **separation of duties** is a means of ensuring that a different team writes the software than the one that installs it. This is done as a measure to help prevent fraud and error. In many companies, the development teams write the code, and the operations teams are responsible for the installation of the software.

A **server** is a device that typically responds to requests from a client device. The server can host a website or provide the business logic; further, a server could store data, send emails, and perform many other tasks. An application will usually use multiple servers, each responsible for a different task (see also **client**).

A **service** is often a software program running continuously on a computer. It runs quietly in the background and responds once a call to the service is made. It can also respond to requests such as generating a PDF file or sending thousands of newsletters.

A **Service-Level Agreement (SLA)** is a type of contract used to define the agreed terms between the provider and the recipient of a service. For software applications, this agreement will contain aspects such as availability, service hours, and responsibilities.

Source code (see **code**).

The **state** is the term used to describe the current circumstances of the running software application. At any given time, one user may have added products to the shopping cart, while another might be in the middle of a credit card payment, and yet other users might be searching for products, and so on.

Stateless, in communication between two systems or between a person and a system, means that no information is temporarily held about the other. Each request and response is handled individually with no prior knowledge of previous communication held in memory.

The **stream** is the continuous flow and display of data before the whole file has loaded, and this is commonly found in music and video. Streaming technology can even be used for real-time broadcasts around the world. The terms *upstream* and *downstream* are used to reference the direction of data flow or other types of directional dependencies, and the difference between the two is simply a matter of perspective.

Supercomputers are those with much better performance than the average computer used at home or at work. A modern smartphone, however, can perform more calculations per second than a supercomputer could around 40 years ago. Therefore, the definition and performance of a supercomputer changes as technology develops.

A **synchronous call/request** is a type of request for data or a service that works well when the response is expected to be fast because no other task will be performed during the waiting time (see also **asynchronous call/request**).

Synthetic monitoring serves to continuously measure how the system is running by simulating real user behaviors.

A **system** is typically a group of components combined to form a single unit.

The **system landscape** is the collection of the company's IT applications and systems. An overview diagram depicts the layout of systems and how they are connected.

The **system layout** generally refers to the way a system and its components are arranged. It is often represented visually to assist in discussions on the layout.

Technical debt is a metaphor in software development. It describes code that was written "quick and dirty" and hence is not futureproof. The code could be inflexible and/or cause a lot of errors. This badly written code will cause additional work and is like accumulated interest that must be paid off over time.

Test coverage is a term used to express in percent what proportion of the code is being tested.

Test-driven development (TDD) involves writing code tests before the actual code is written. To a layperson, this may sound like the wrong way to approach software development, but determining what the code should and shouldn't do first has proven very effective.

The **testing** process verifies if the software fulfills the requirements that have been defined. Over a hundred different types of testing are available for software projects.

In **user acceptance testing (UAT)**, end users test the software application. The UAT is typically one of the very last test phases done before the changes are launched in production.

A **ubiquitous language** helps in defining a common language and set of terms for a project, for a business area, or even company-wide. This common language of agreed-upon terms enables all the members of different teams to communicate clearly. One important goal is to avoid any misunderstandings that can arise due to certain words and terms having more than one meaning.

A **unified modeling language (UML)** is used to visualize a system design. It is most often done prior to the programming but can also be done later to visualize an existing system.

The term **usability** refers to the degree of intuitive navigation and easy interaction in a piece of software. The user should be able to achieve their goal in as few steps as possible.

A **use case** is a formal list of actions required for a goal to be achieved by a user. Typically, a use case will include many details, including both the successful and unsuccessful routes. The use case is a more formal notation than a user story. A use case diagram is used to visualize such an action done by a user.

The **user** is anyone using a system or software application.

The **user experience (UX)** is concerned with how a user feels when using the software.

A **user interface (UI)** is the part of the software users can see and interact with. The technical name for this visual part of the software is the *graphical user interface (GUI)*, which is a subset of the UI. The UI also covers non-graphical interfaces, like voicemail, card readers, and so forth. The GUI will typically consist of a color screen but could also be a monochrome screen on a printer or a simple display on a home heating system.

User stories are brief and informal descriptions of users' goals. A user story is a tool used in Agile software development to capture a description of a software feature from an end user perspective. Often these are assembled in group sessions. Each user story is written on a paper card. The user story describes the type of user, what they want, and why. A user story helps create a simplified description of a requirement.

Version/versioning can be a number or a name that helps identify the software version.

A **virtual machine** is the term used to mean an emulation of a computer. The machine no longer runs on hardware specifically for this device and operating system but sits on a larger system that may host dozens of other virtual computers.

Virtualization refers to the processes of creating virtual machines, virtual hardware, or other types of emulations.

The **Waterfall methodology (plan-driven approach)** is the name given to the classic project method where the steps, such as analysis, design, and implementation, are conducted in sequential order. This methodology works well for projects with stable requirements.

A **web server** is a device that responds to requests typically made through a browser by a human user.

A **wiki** is a system for sharing knowledge with others.

Wireframes are created to give the users and stakeholders a first impression of the user interface. This could be the visual representation of a smartphone app or the display of a washing machine, a website, or a desktop application. The wireframe shows which visual elements go where. The wireframe can also help the business, designers, and developers to consider any improvements in the design before it is programmed (see also **mockup**).

Work in progress (WIP) is a task that is currently being carried out and is not yet completed.

A **workstation** is a powerful computer, set up with the tools and rights required for a developer.

YAGNI: *You aren't gonna need it* refers to a feature or part of software that is not necessary now (but may be needed in the future). In Agile projects especially, the goal is to build only the software that is currently essential, leaving aside anything else that is based on assumptions.

References and Further Reading

The books, publications, and further references listed in this section are all chosen for their value as excellent materials to learn from. All of the sources are highly regarded and have fantastic ratings. I can strongly recommend every one of these resources.

Those highlighted in gray are sources that contain a high level of technical content and may be difficult for non-techies to follow. The other materials are suitable for a wider audience.

Referenced Books and Publications

[Alexander et al. 1977]

Alexander, C., Ishikawa, S., Silverstein, M., Jacobsen, M., Fiksdahl-King, I. & Shlomo, A. (1977). *A Pattern Language: Towns, Buildings, Construction.* New York, NY: Oxford University Press.

[Bodell 2017]

Bodell, L. (2017). *Why Simple Wins: Escape the Complexity Trap and Get to Work That Matters.* New York, NY: Bibliomotion, Inc.

© Jonathan Peter Crosby 2023
J. P. Crosby, *The Business Manager's Guide to Software Projects,*
https://doi.org/10.1007/978-1-4842-9231-0

[Bonner & Bolinger 2013]

Bonner, B. L., & Bolinger, A. R. (2013). *Separating the confident from the correct: Leveraging member knowledge in groups to improve decision-making and performance.* Department of Management, The University of Utah: Organizational Behavior and Human Decision Processes, Vol. 122, pp. 214–221. Discipline based–refereed, published, 07/2013.

[Brooks 1995]

Brooks, F. P., Jr. (1995). *The Mythical Man-Month: Essays on Software Engineering* (Anniversary ed.). Reading, MA: Addison-Wesley.

[Burke 1945]

Burke, K. (1945). *A Grammar of Motives.* Berkeley and Los Angeles, California. University of California Press, p. 503.

[DeMarco & Lister 2013]

DeMarco, T., & Lister, T. R. (2013). *Peopleware: Productive Projects and Teams* (3rd ed.). Reading, MA: Addison-Wesley.

[Evans 2004]

Evans, E. (2004). *Domain-Driven Design: Tackling Complexity in the Heart of Software* (1st ed.). Reading, MA: Addison-Wesley.

[Garrett et al. 2012]

Garrett, D. (2012). *Project Pain Reliever: A Just-in-Time Handbook for Anyone Managing Projects.* Fort Lauderdale, FL: J. Ross Publishing.

[Glass & Holyoak 1986]

Glass, A. L., & Holyoak, K. J. (1986). *Cognition* (2nd ed.). Reading, MA: Addison-Wesley.

[Horowitz 2014]

Horowitz, B. (2014). *The Hard Thing About Hard Things: Building a Business When There Are No Easy Answers.* New York, NY: HarperCollins Publishers.

[Hunt & Thomas 1999]

Hunt, A., & Thomas, D. (1999). *The Pragmatic Programmer: From Journeyman to Master.* Reading, MA: Addison-Wesley.

[Kano et al. 1984]

Kano, N., Seraku, N., Takahashi, F., & Tsuji, S. (1984). *Attractive quality and must-be quality* (in Japanese). Hinshitsu, The Journal of the Japanese Society for Quality Control, April, pp. 39–48.

[Kim, Behr & Spafford 2013]

Kim, G., Behr, K., & Spafford G. (2013). *The Phoenix Project: A Novel About IT, DevOps, and Helping Your Business Win* (Revised ed.). Portland, OR: IT Revolution Press.

[Kosslyn 1988]

Kosslyn, S. M. (1988). *Imagery in learning: perspectives in memory research.* (Ed. M. S. Gazzaniga). Cambridge, MA: MIT Press.

[Lencioni 2002]

Lencioni, P. M. (2002). *The Five Dysfunctions of a Team.* San Francisco, CA: Jossey-Bass.

[McConnell 2004]

McConnell, S. (2004). *Code Complete: A Practical Handbook of Software Construction* (2nd ed.). Redmond, WA: Microsoft Press.

[Newport 2016]

Newport, C. (2016). *Deep Work: Rules for Focused Success in a Distracted World.* New York, NY: Hachette Book Group, Inc.

[Peters 2013]

Peters, S. (2013). *The Chimp Paradox: The Mind Management Program to Help You Achieve Success, Confidence and Happiness.* New York, NY: Penguin Group (USA) Inc.

[Robertson & Robertson 2013]

Robertson, S., & Robertson, J. (2013). *Mastering the Requirements Process: Getting Requirements Right* (3rd ed.). Reading, MA: Addison-Wesley.

[Tuckman 1965]

Tuckman, B. W. (1965). *Developmental sequence in small groups.* Psychological Bulletin, Vol. 63, pp. 384–399.

Referenced Online Sources

As the URLs can change over time, I have included the links on my website and will aim to keep these up to date: www.SoftwareGuide.blog.

[Ackmann 2016]

Ackmann, A. (March 3, 2016). *Writing in the workplace: email, memos, reports, and social messaging.* Retrieved from www.pluralsight.com/courses/writing-workplace-email-memos-reports-social

[Bradd 2016]

Bradd, S. (January 14, 2016). *Visual storytelling: finding common ground.* Retrieved from http://drawingchange.com/visualstorytelling-finding-common-ground-and-building-power

[British Airways Flight Chaos 2017]

British Airways flight chaos lessens after weekend of disruption (May 29, 2017). Retrieved from www.bbc.com/news/ uk-40081112

[Bruns, Eschenbach & Maier 2016]

Bruns, W., Eschenbach, S., & Maier, E. (2016). *Experience–the invisible success factor in commercial enterprises* (in German). Retrieved from www.rfh-koeln.de/sites/rfh_koelnDE/myzms/content/e380/e1184/e29466/e32283/e32285/StudieMETIS_Erfahrungderunsichtbare ErfolgsfaktorinWirtschaftsunternehmen_ger.pdf

[Buchanan]

Buchanan, I. (n.d.). *Agile and DevOps: Friends or Foes?* Retrieved from www.atlassian.com/agile/devops

[Canny 2018]

Canny, W. (April 19, 2018). *Deutsche Bank's Bad News Gets Worse With $35 Billion Flub.* Retrieved from www.bloomberg.com/news/articles/2018-04-19/deutsche-bank-flub-said-to-send-35-billion-briefly-out-the-door

[Daemon 2017]

Daemon, K. (March 3, 2017). *Why agile companies earn more* (in German). Retrieved from `http://app.wiwo.de/erfolg/ unternehmensfuehrung/management-ansaetze-warumagile- unternehmen-mehr-verdienen/19513702.html`

[Dorer 2017]

Dorer, B. (June 20, 2017). *Glassdoor: Clorox CEO Benno Dorer #1 Highest Rated CEO 2017*. Retrieved from `www.youtube.com/ watch?v=K6GDaPh1Ogk`

[Edmondson 2014]

Edmondson, A. (May 4, 2014). *Building a psychologically safe workplace: Amy Edmondson at TEDxHGSE*. Retrieved from `www.youtube. com/watch?v=LhoLuui9gX8`

[Feasibility Studies 2017]

Feasibility studies for construction projects (2017). Retrieved from `www.designingbuildings.co.uk/wiki/Feasibility_studies_for_ construction_projects`

[Forrester Consulting 2013]

Continuous Delivery: A Maturity Assessment Model. Building Competitive Advantage With Software Through A Continuous Delivery Process (March 2013). Cambridge, MA. Forrester Research, Inc. Retrieved from `http://info.thoughtworks.com/rs/thoughtworks2/images/ Continuous%20Delivery%20_%20A%20Maturity%20Assessment%20 ModelFINAL.pdf`

[Fowler Chad 2014]

Fowler, C. (January 20, 2014). *Your most important skill: empathy.* Retrieved from `http://chadfowler.com/2014/01/19/empathy.html`

[Fowler Martin 2003]

Fowler, M. (2003). *TechnicalDebt*. Retrieved from `https:// martinfowler.com/bliki/TechnicalDebt.html`

[Fuda 2012]

Fuda, P. (November 11, 2012). *3 reasons why metaphors are powerful.* Retrieved from www.peterfuda.com/2012/11/06/3-reasons why-metaphors-are-powerful

[Grant 2016]

Grant, A. (November 2016). *Are you a Giver or a Taker?* Retrieved from www.ted.com/talks/adam_grant_are_you_a_giver_or_a_taker

[Hanselman 2013]

Hanselman, S. (September 29, 2013). *The Myth of the Rockstar Programmer.* Retrieved from www.hanselman.com/blog/TheMythOfTheRockstarProgrammer.aspx

[High-Level Design–Wikipedia]

High-Level Design (n.d.). In Wikipedia. Retrieved from https://en.wikipedia.org/wiki/High-level_design

[Hohpe 2016]

Hohpe, G. (July 5, 2016). *37 Things or "Where have all my ramblings gone?"* Retrieved from www.enterpriseintegrationpatterns.com/ramblings/94_37things.html

[Kassel–Wikivoyage]

Kassel-Wilhelmshöhe Bahnhof (n.d.) (in German). In Wikivoyage. Retrieved from https://de.wikivoyage.org/wiki/Kassel-Wilhelmsh%C3%B6he_Bahnhof

[Manifesto for Agile Software Development 2001]

Beck, K., Beedle, M., van Bennekum, A., Cockburn, A., Cunningham, W., Fowler, M., Grenning, J., Highsmith, J., Hunt, A., Jeffries, R., Kern, J., Marick, B., Martin, R. C., Mallor, S., Shwaber, K., & Sutherland, J. (2001). *Manifesto for Agile software development.* Retrieved from http://agilemanifesto.org

[Martin 2014]

Martin, B. (November 11, 2014). *Uncle Bob Martin.* Retrieved from www.youtube.com/watch?v=QHnLmvDxGTY

[Nielsen, Niu & Meng 2016]

Nielsen, C., Niu, D., & Meng, S. (November 7, 2016). *Measuring Your Employees' Invisible Forms of Influence*. Retrieved from `https://hbr.org/2016/11/measuring-your-employees-invisibleforms-of-influence`

[Nottaris et al. 2016]

Nottaris, M., Nufer, P., Günther, S., Boris Gygax, B., & Messmer, L. (2016). *Einstein: Die gigantischste Zugfabrik der Welt*. Retrieved from `www.srf.ch/sendungen/einstein/china-dieneue-bahnnation`

[Owens 2013]

Owens, K. (January 21, 2013). *Active Listening: Katie Owens at TEDxYouth@Conejo*. Retrieved from `www.youtube.com/watch?v=WER63AY8zB8`

[Seewald et al. 2018]

Seewald, C., Frei, K., Hille, S., & Wick, H. (2018). *Einstein: Sandra Akmansoy, Herrin der Superprojekte*. Retrieved from `www.srf.ch/sendungen/einstein/grossbaustellen-und-milliardenprojekte`

[Shore 2004]

Shore, J. (September/October 2004). *Fail Fast*. Retrieved from `https://martinfowler.com/ieeeSoftware/failFast.pdf`

[Spolsky 2006]

Spolsky, J. (August 9, 2006). *The Econ 101 Management Method*. Retrieved from `www.joelonsoftware.com/2006/08/09/the-econ-101-management-method/`

[Spotify Engineering Culture 2014]

Kniberg, H. (2014). *Spotify engineering culture*. Retrieved from `https://labs.spotify.com/2014/03/27/spotify-engineering-culture-part-1` and `https://labs.spotify.com/2014/09/20/spotify-engineering-culture-part-2`

[UNC Computer Science 2015]

UNC Computer Science (June 6, 2015). *Frederick P. Brooks Jr.–Last Blast*. Retrieved from `www.youtube.com/watch?v=zJ594Pl6dQc`

[Wigglesworth, Hughes & Lockett 2017]

Wigglesworth, R., Hughes, J., & Lockett, H. (July 4, 2017). In the Financial Times. Retrieved from www.ft.com/content/fbb44c3e-6053-1 1e7-91a7-502f7ee26895

[Wonders of the World 2015]

Wonders of the world (June 11, 2015). Man Made Marvels, Discovery Channel. *Sydney Opera House–documentary*. Retrieved from www.youtube.com/watch?v=EXBCaGbOdy8

Further Reading

Here are some great videos to help you dig a little deeper.

Martin Fowler–Agile Essence and Fluency: www.youtube.com/watch?v=URlnxbaHhTs

The beauty of data visualization by David McCandless: www.youtube.com/watch?v=5Zg-C8AAIGg

What is DevOps? In Simple English: www.youtube.com/watch?v=_ I94-tJlo

Index

A

A/B tests, 207
Acceptance criteria, 75, 105, 156, 157, 163, 181, 188, 204
Accountability, 100, 124, 145
Ackmann, Alan, 139
Acoustic ceiling, 54
Acquisitions, 101
Age diversity, 92
Agile approach, 13
Agile Manifesto, 88
Agile methodology
    about, 88, 108
    benefits, 61
    and iterations, 13
    Manifesto, 88
    and requirements, 157, 158
    terminology, 198
    and testing, 208
Agility, 80, 102, 253
Akmansoy, Sandra, 152
Alexander, Christopher, 24, 191
Alibaba, 82
Analysis paralysis, 157
Analytics, 182, 216, 217
Architects, 3–5, 20, 23, 24, 28, 36–39, 61, 100, 106, 114, 146, 175, 176, 181

Assumptions, in communications, 111, 117
Asynchronous, 238, 239
Audits, 137, 173
Authentication, 237
Authorization, 235, 237
Automation, 95, 97, 98, 127, 180, 207, 208
Awesome, definition of, 153

B

Back-office division, 80
Bandwidth, 253, 254, 266
Barclays Bank, 35
Beauty *vs.* practicality, 22, 24
Beck, Harry, 83, 85, 251
Behavior (of application), 190
Big-bang approach, 12, 218, *See also* Waterfall methodology
Bits, 4, 241
Black box, 182
Bodell, Lisa, 51, 127, 145
Bottlenecks, 206, 233
Boundary case testing, 210
Brainstorming, 153
Breaking changes, 196
Brooks, Fred P., 93, 106, 122, 151, 161
    estimates, 48, 158, 177, 178, 183

© Jonathan Peter Crosby 2023
J. P. Crosby, *The Business Manager's Guide to Software Projects,*
https://doi.org/10.1007/978-1-4842-9231-0

D